Topics on TOURNAMENTS in GRAPH THEORY

John W. Moon
University of Alberta

Dover Publications, Inc.
Mineola, New York

Bibliographical Note

Topics on Tournaments in Graph Theory, first published by Dover Publications, Inc., in 2015, is an unabridged republication of *Topics on Tournaments*, originally published in 1968 by Holt, Rinehart and Winston, Inc., New York, as part of the "Athena Series: Selected Topics in Mathematics." This edition is published by special arrangement with Cengage Learning, Inc., Belmont, California.

Library of Congress Cataloging-in-Publication Data

Moon, John W.
    Topics on tournaments in graph theory / John W. Moon. — Dover edition.
        pages cm
    Originally published: New York : Holt, Rinehart and Winston, 1968.
    Includes bibliographical references and index.
    ISBN-13: 978-0-486-79683-3
    ISBN-10: 0-486-79683-3
    1. Tournaments (Graph theory) I. Title.
    QA166.M66 2015
    511'.54—dc23

                                                                    2014043614

Manufactured in the United States by Courier Corporation
79683301   2015
www.doverpublications.com

Preface

My object has been to collect in one volume various results on tournaments that are scattered throughout the mathematical literature. The material here is quite elementary for the most part, and any reader who is familiar with the elements of what is nowadays called finite mathematics should have little difficulty in understanding most of the theorems. I have tried to put what I thought were the most basic definitions and results in the earlier sections; most of the later sections can be read independently of each other. The exercises vary between routine verifications and unsolved problems.

J.W.M.

Edmonton, Alberta
February 1968

Contents

1. Introduction

A (round-robin) *tournament* T_n consists of n nodes $p_1, p_2, \cdots, p_n$ such that
each pair of distinct nodes p_i and p_j is joined by one and only one of the
oriented arcs $\overrightarrow{p_i p_j}$ or $\overrightarrow{p_j p_i}$. If the arc $\overrightarrow{p_i p_j}$ is in T_n, then we say that p_i *dominates*
p_j (symbolically, $p_i \rightarrow p_j$). The relation of dominance thus defined is a
complete, irreflexive, antisymmetric, binary relation. The *score* of p_i is the
number s_i of nodes that p_i dominates. The *score vector* of T_n is the ordered
n-tuple $(s_1, s_2, \cdots, s_n)$. We usually assume that the nodes are labeled
in such a way that $s_1 \leq s_2 \leq \cdots \leq s_n$.

Tournaments provide a model of the statistical technique called the
method of paired comparisons. This method is applied when there are a
number of objects to be judged on the basis of some criterion and it is
impracticable to consider them all simultaneously. The objects are compared two at a time and one member of each pair is chosen. This method and
related topics are discussed in David (1963) and Kendall (1962). Tournaments have also been studied in connection with sociometric relations in
small groups. A survey of some of these investigations is given by Coleman
(1960). Our main object here is to derive various structural and combinatorial properties of tournaments.

Exercises

1. Two tournaments are *isomorphic* if there exists a one-to-one dominance-
preserving correspondence between their nodes. The nonisomorphic
tournaments with three and four nodes are illustrated in Figure 1. Deter-

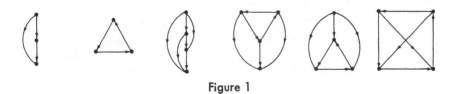

Figure 1

mine the number of ways of assigning the labels to the nodes of these
tournaments and verify that there are a total of $2^{\binom{n}{2}}$ labeled tournaments T_n
when $n = 3, 4$.

2. The *complement* of a tournament is obtained by reversing the orientation
of all its arcs. A tournament is *self-complementary* if it is isomorphic to its

complement. Show that self-complementary tournaments T_n exist if and only if n is odd. [Sachs (1965).]

2.   Irreducible Tournaments

A tournament T_n is *reducible* if it is possible to partition its nodes into two nonempty sets B and A in such a way that all the nodes in B dominate all the nodes in A; the tournament is *irreducible* if this is not possible. It is very easy to determine whether a tournament T_n is reducible; if $(s_1, s_2, \cdots, s_n)$ is the score vector of T_n and $s_1 \leq s_2 \leq \cdots \leq s_n$, then T_n is reducible if and only if the equation

$$\sum_{i=1}^{k} s_i = \binom{k}{2}$$

holds for some value of k less than n.

The (dominance) *matrix of the tournament* T_n is the n by n matrix $M(T_n) = [a_{ij}]$ in which a_{ij} is 1 if $p_i \rightarrow p_j$ and 0 otherwise. All the diagonal entries are 0. A tournament matrix satisfies the equation

$$M + M^T = J - I,$$

where J is the matrix of 1's and I is the identity matrix. If the tournament T_n is reducible and the scores $s_i = \Sigma_{j=1}^{n} a_{ij}$ are in nondecreasing order, then its matrix has the structure

$$M(T_n) = \left|\begin{array}{c|c} M_1 & 0 \\ \hline 1 & M_2 \end{array}\right|,$$

where M_1 and M_2 are the matrices of the tournaments defined by the sets A and B of the preceding paragraph.

There are $2^{\binom{n}{2}}$ labeled tournaments T_n. We now derive an approximation for $P(n)$, the probability that a random tournament T_n is irreducible.

Every reducible tournament T_n has a unique decomposition into irreducible subtournaments $T^{(1)}, T^{(2)}, \cdots, T^{(l)}$ such that every node in $T^{(j)}$ dominates every node in $T^{(i)}$ if $1 \leq i < j \leq l$. The probability that $T^{(1)}$ has t nodes is

$$\frac{\binom{n}{t} P(t)\, 2^{\binom{t}{2}}\, 2^{\binom{n-t}{2}}}{2^{\binom{n}{2}}} = \binom{n}{t} P(t) \left(\frac{1}{2}\right)^{t(n-t)}.$$

For each of the $\binom{n}{t}$ subsets of t nodes, there are $P(t)\, 2^{\binom{t}{2}}$ possible choices for $T^{(1)}$; the $\binom{n-t}{2}$ arcs joining the remaining $n-t$ nodes may be oriented arbitrarily.

It is possible for t to be any positive integer less than n. It follows that

$$P(n) = 1 - \sum_{t=1}^{n-1} \binom{n}{t} P(t) \left(\frac{1}{2}\right)^{t(n-t)}, \tag{1}$$

since these cases are mutually exclusive. R. A. MacLeod used this formula to calculate the first few values of $P(n)$ given in Table 1.

Table 1. $P(n)$, the probability that a tournament T_n is irreducible.

n	1	2	3	4	5	6	7	8	9
$P(n)$	1	0	.25	.375	.53125	.681152	.799889	.881115	.931702

n	10	11	12	13	14	15	16
$P(n)$	.961589	.978720	.988343	.993671	.996587	.998171	.999024

The following bound is a stronger form of a result due to Moon and Moser (1962b).

THEOREM 1. If $Q(n)$ denotes the probability that a random tournament T_n is reducible, then

$$\left| Q(n) - \frac{n}{2^{n-2}} \right| < \frac{1}{2}\left(\frac{n}{2^{n-2}}\right)^2 \qquad \text{if } n \geq 2.$$

Proof. It follows from (1) that

$$P(n) > 1 - 2n\left(\frac{1}{2}\right)^{n-1} - \binom{n}{2}\left(\frac{1}{2}\right)^{2(n-2)} - \sum_{t=3}^{n-3} \binom{n}{t}\left(\frac{1}{2}\right)^{t(n-t)}.$$

The terms in the sum are largest for the extreme values of t. Consequently,

$$P(n) > 1 - n(\tfrac{1}{2})^{n-2} - n^2(\tfrac{1}{2})^{2n-3} + \left(n(\tfrac{1}{2})^{2n-3} - (n-5)\binom{n}{3}(\tfrac{1}{2})^{3n-9}\right).$$

If $n \geq 14$, then

$$P(n) > 1 - n(\tfrac{1}{2})^{n-2} - n^2(\tfrac{1}{2})^{2n-3}, \tag{2}$$

since the expression within the parenthesis is positive.

To obtain an upper bound for $P(n)$, we retain only the three largest terms in the sum in (1). Therefore,

$$P(n) < 1 - n(\tfrac{1}{2})^{n-1} - nP(n-1)(\tfrac{1}{2})^{n-1} - \binom{n}{2}P(n-2)(\tfrac{1}{2})^{2(n-2)}.$$

If $n - 2 \geq 14$, we can use inequality (2) to bound $P(n - 1)$ and $P(n - 2)$ from below; the resulting expression may be simplified to yield the inequality

$$P(n) < 1 - n(\tfrac{1}{2})^{n-2} + n^2(\tfrac{1}{2})^{2n-3}. \tag{3}$$

Theorem 1 now follows from (2), (3), and the data in Table 1.

Exercises

1. Verify that $P(3) = \tfrac{1}{4}$ and $P(4) = \tfrac{3}{8}$ by examining the tournaments in Figure 1.

2. Deduce inequality (3) from the two preceding inequalities.

3. Prove that T_n is irreducible if the difference between every two scores in T_n is less than $\tfrac{1}{2}\, n$. [L. Moser.]

4. Let the score vector of T_n be $(s_1, s_2, \cdots, s_n)$, in nondecreasing order. Show that p_i and p_j are in the same irreducible subtournament of T_n if $0 \leq s_j - s_i < (j - i + 1)/2$. [L. W. Beineke.]

5. If $T(x) = \Sigma_{n=1}^{\infty} 2^{\binom{n}{2}} x^n/n!$ and $t(x) = \Sigma_{n=1}^{\infty} P(n) 2^{\binom{n}{2}} x^n/n!$, then show that $t(x) = T(x)/[1 + T(x)]$. (These are only formal generating functions, so questions of convergence may be ignored.)

6. Let $T_n - p_i$ denote the tournament obtained from T_n by removing the node p_i (and all arcs incident with p_i). If T_n and H_n are two tournaments with $n(n \geq 5)$ nodes p_i and q_i, respectively, such that $T_n - p_i$ is isomorphic to $H_n - q_i$ for all i, then is T_n necessarily isomorphic to H_n? (Consider first the case in which T_n and H_n are reducible.) Consider the analogous problem when arcs instead of nodes are removed from T_n and H_n. [See Harary and Palmer (1967).]

3.   Strong Tournaments

For any subset X of the nodes of a tournament T_n, let

$$\Gamma(X) = \{q : p \to q \text{ for some } p \in X\},$$

and, more generally, let

$$\Gamma^m(X) = \Gamma(\Gamma^{m-1}(x)), \qquad \text{for } m = 2, 3, \cdots.$$

Notice that T_n is reducible if and only if $\Gamma(X) \subseteq X$ for some nonempty proper subset X of the nodes. A tournament T_n is *strongly connected* or *strong* if and only if for every node p of T_n the set

$$\{p\} \cup \Gamma(p) \cup \Gamma^2(p) \cup \cdots \cup \Gamma^{n-1}(p)$$

contains every node of T_n. The following theorem apparently appeared first in a paper by Rado (1943); it was also found by Roy (1958) and others.

THEOREM 2. A tournament T_n is strong if and only if it is irreducible.

Proof. If T_n is reducible, then it is obviously not strong. If T_n is not strong, then for some node p the set

$$A = \{p\} \cup \Gamma(p) \cup \Gamma^2(p) \cup \cdots \cup \Gamma^{n-1}(p)$$

does not include all the nodes of T_n. But then all the nodes not in A must dominate all the nodes in A; consequently T_n is reducible.

In view of the equivalence between strong connectedness and irreducibility, it is a simple matter to determine whether or not a given tournament is strongly connected by using the observation in the introduction to Section 2. For types of graphs other than tournaments in which, for example, not every pair of nodes is joined by an arc, it is sometimes convenient to use the properties of matrix multiplication to deal with problems of connectedness.

Exercises

1. Show that the tournament T_n is strong if and only if all the entries in the matrix

$$M(T_n) \dotplus M^2(T_n) \dotplus \cdots \dotplus M^{n-1}(T_n)$$

are 1s. (Boolean addition is to be used here, that is, $1 \dotplus 1 = 1 \dotplus 0 = 1$ and $0 \dotplus 0 = 0$.) [Roy (1959) and others.]

2. Prove that, in any tournament T_n, there exists at least one node p such that the set

$$\{p\} \cup \Gamma(p) \cup \Gamma^2(p) \cup \cdots \cup \Gamma^{n-1}(p)$$

contains every node of T_n. [See Bednarek and Wallace (1966) for a relation-theoretic extension of this result.]

4.   Cycles in a Tournament

A sequence of arcs of the type $\overrightarrow{ab}, \overrightarrow{bc}, \cdots, \overrightarrow{pq}$ determines a *path* $P(a,q)$ from a to q. We assume that the nodes $a, b, c, \cdots, q$ are all different. If the arc $\overrightarrow{qa}$ is in the tournament, then the arcs in $P(a, q)$ plus the arc $\overrightarrow{qa}$ determine a *cycle*. The *length* of a path or a cycle is the number of arcs it contains. A *spanning* path or cycle is one that passes through every node in a tournament. A tournament is strong if and only if each pair of nodes is contained in some cycle.

Moser and Harary (1966) proved that an irreducible tournament T_n contains a k-cycle (a cycle of length k) for $k = 3, 4, \cdots, n$. [Their argument was a refinement of the argument Camion (1959) used to prove that a tournament T_n ($n \geq 3$) contains a spanning cycle if and only if it is irreducible.] The following slightly stronger result is proved in essentially the same way.

THEOREM 3. Each node of an irreducible tournament T_n is contained in some k-cycle, for $k = 3, 4, \cdots, n$.

Proof. Let a be an arbitrary node of the irreducible tournament T_n. There must be some arc $\overrightarrow{pq}$ in T_n where $q \rightarrow a$ and $a \rightarrow p$; otherwise T_n would be reducible. Consequently, node a is contained in some 3-cycle. Let

$$C = \{\; \overrightarrow{ab}, \overrightarrow{bc}, \cdots, \overrightarrow{lm}, \overrightarrow{ma}, \}$$

be a k-cycle containing the node a, where $3 \leq k < n$. We shall show that there also exists a $(k + 1)$-cycle containing a.

We first suppose there exists some node p not in the cycle such that p both dominates and is dominated by nodes that are in the cycle. Then there must be two consecutive nodes of the cycle, e and f say, such that $e \rightarrow p$ and $p \rightarrow f$. We can construct a $(k + 1)$-cycle containing node a simply by replacing the arc $\overrightarrow{ef}$ of C by the arcs $\overrightarrow{ep}$ and $\overrightarrow{pf}$. (This case is illustrated in Figure 2(a).)

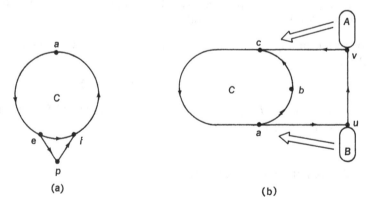

(a)   (b)

Figure 2

Now let A and B denote, respectively, the sets of nodes of T_n not in cycle C that dominate, or are dominated by, every node of C. We may assume that every node of T_n not in C belongs either to A or to B. Since T_n is irreducible, both A and B must be nonempty and some node u of B

must dominate some node v of A. But then we can construct a $(k + 1)$-cycle containing node a by replacing the arcs $\overrightarrow{ab}$ and $\overrightarrow{bc}$ of C by the arcs $\overrightarrow{au}$, $\overrightarrow{uv}$, and $\overrightarrow{vc}$. (This case is illustrated in Figure 2(b).) This completes the proof of the theorem by induction.

Exercises

1. Examine the argument Foulkes (1960) gave to show that an irreducible tournament has a spanning cycle. [See also Fernández de Trocóniz (1966).]

2. Let us say that a tournament has property P_k if every subset of k nodes determines at least one k-cycle. Show that T_n has a spanning cycle if it has property P_k for some k such that $3 \leq k \leq n$.

3. What is the maximum number of arcs $\overrightarrow{pq}$ that an irreducible tournament T_n can have such that, if the arc $\overrightarrow{pq}$ is replaced by the arc $\overrightarrow{qp}$, then the resulting tournament is reducible?

4. What is the least integer $r = r(n)$ such that any irreducible tournament T_n can be transformed into a reducible tournament by reversing the orientation of at most r arcs?

5. A tournament T_r is a *subtournament* of a tournament T_n if there exists a one-to-one mapping f between the nodes of T_r and a subset of the nodes of T_n such that, if $p \rightarrow q$ in T_r, then $f(p) \rightarrow f(q)$ in T_n. Let T_r $(r > 1)$ denote an irreducible subtournament of an irreducible tournament T_n. Prove that there exist k-cycles C such that every node of T_r belongs to C for $k = r$, $r + 1, \cdots, n$ with the possible exception of $k = r + 1$. Characterize the exceptional cases.

6. Let T_r $(r > 1)$ denote a reducible subtournament of an irreducible tournament T_n. Determine bounds for $h(T_n)$, the least integer h for which there exists an h-cycle C that contains every node of T_r.

7. A *regular* tournament is one in which the scores of the nodes are all as nearly equal as possible. Let T_r $(r > 1)$ denote a subtournament of a regular tournament T_n. Prove that there exist k-cycles C such that every node of T_r belongs to C for $k = r, r + 1, \cdots, n$ with the possible exception of $k = r$ or $k = r + 1$. [See Kotzig (1966).]

8. Prove that every arc of a regular tournament T_n with an odd number of nodes is contained in a k-cycle, for $k = 3, 4, \cdots, n$. [Alspach (1967).]

9. P. Kelly has asked the following question: Is it true that the arcs of every regular tournament T_n with an odd number of nodes can be partitioned into $\frac{1}{2}(n - 1)$ arc-disjoint spanning cycles?

5.   Strong Subtournaments of a Tournament

Let $S(n, k)$ denote the maximum number of strong subtournaments T_k that can be contained in a tournament T_n ($3 \leq k \leq n$). The following result is due to Beineke and Harary (1965).

THEOREM 4.   If $[x]$ denotes the greatest integer not exceeding x, then

$$S(n, k) = \binom{n}{k} - \left[\frac{1}{2}(n+1)\right]\binom{[\frac{1}{2}n]}{k-1} - \left[\frac{1}{2}n\right]\binom{[\frac{1}{2}(n-1)]}{k-1}.$$

Proof. Let $(s_1, s_2, \cdots, s_n)$ be the score vector of a tournament T_n. The number of strong subtournaments T_k in T_n certainly cannot exceed

$$\binom{n}{k} - \sum_{i=1}^{n}\binom{s_i}{k-1}.$$

This is because the terms subtracted from $\binom{n}{k}$, the total number of sub-tournaments T_k in T_n, enumerate those subtournaments T_k in which one node dominates all the remaining $k - 1$ nodes; such tournaments are certainly not strong. It is a simple exercise to show that the sum attains its minimum value when the s_i's are as nearly equal as possible. Consequently,

$$S(n, k) \leq \binom{n}{k} - \left[\frac{1}{2}(n+1)\right]\binom{[\frac{1}{2}n]}{k-1} - \left[\frac{1}{2}n\right]\binom{[\frac{1}{2}(n-1)]}{k-1}.$$

To show that equality actually holds, it suffices to exhibit a regular tournament R_n with the following property.

(a) Every subtournament T_k of R_n is either strong or has one node that dominates all the remaining $k - 1$ nodes.

If n is odd, let R_n denote the regular tournament in which $p_i \rightarrow p_j$ if and only if $0 < j - i \leq (n - 1)/2$ (the subtraction is modulo n). We shall show that R_n has the following property.

(b) Every subtournament T_k of R_n either is strong or has no cycles.

Property (b) holds for any tournament when $k = 3$. We next show that it holds for R_n when $k = 4$. If it did not, then R_n would contain a subtournament T_4 consisting of a 3-cycle and an additional node that either dominates every node of the cycle or is dominated by every node of the cycle. We may suppose that the former is the case without loss of generality. Let p_1, p_i, and p_j be the nodes of the cycle, proceeding according to its orientation, and let p_k be the fourth node of T_4. Then $i \leq 1 + \frac{1}{2}(n - 1)$, $j \leq i + \frac{1}{2}(n - 1)$, $j > 1 + \frac{1}{2}(n - 1)$, and $n \geq k > i + \frac{1}{2}(n - 1)$, from the definition of

R_n. Hence, $n \geq k > j > \frac{1}{2}(n + 1)$, and p_j must dominate p_k in R_n and T_4. This contradiction shows that R_n satisfies (b) when $k = 4$.

If n is even, let R_n denote the regular tournament in which $p_i \rightarrow p_j$ if and only if $0 < j - i \leq \frac{1}{2}n$ (the subtraction is modulo $n + 1$). This tournament satisfies (b) when $k = 3$ or 4, since it is a subtournament of the tournament R_{n+1} defined earlier with an odd number of nodes.

Let T_k be any subtournament with k nodes of R_n ($k > 4$). If T_k has any cycles at all, then it must have a 3-cycle C. If T_k is not strong, then there must be a node p that either dominates every node of C or is dominated by every node of C. In either case, this would imply the existence of a subtournament T_4 in R_n that is not strong but which has a cycle. This contradicts the result just established. Hence, every subtournament T_k of R_n is either strong or has no cycles. We have shown that the regular tournament R_n has Property (b). It is easy to show that Property (b) implies Property (a), so the theorem is now proved.

There are only two different types of tournaments T_3 (see Figure 1); it follows that equality holds in the second statement in the proof of Theorem 4 when $k = 3$. This implies the following result, found by Kendall and Babington Smith (1940), Szele (1943), Clark [see Gale (1964)], and others.

COROLLARY. Let $c_3(T_n)$ denote the number of 3-cycles in the tournament T_n. If $(s_1, s_2, \cdots, s_n)$ is the score vector of T_n, then

$$c_3(T_n) = \binom{n}{3} - \sum_{i=1}^{n} \binom{s_i}{2} \leq \begin{cases} \frac{1}{24}(n^3 - n) & \text{if } n \text{ is odd,} \\ \frac{1}{24}(n^3 - 4n) & \text{if } n \text{ is even.} \end{cases}$$

Equality holds throughout only for regular tournaments.

The next corollary follows from the observation that every strong tournament T_4 has exactly one 4-cycle.

COROLLARY. The maximum number of 4-cycles possible in a tournament T_n is

$$S(n, 4) = \begin{cases} \frac{1}{48} n(n + 1)(n - 1)(n - 3) & \text{if } n \text{ is odd,} \\ \frac{1}{48} n(n + 2)(n - 2)(n - 3) & \text{if } n \text{ is even.} \end{cases}$$

Colombo (1964) proved this result first by a different argument. It seems to be difficult to obtain corresponding results for cycles of length greater than four. (See Exercise 3 at the end of Section 10.)

The following result is proved by summing the equation in Theorem 4 over k.

COROLLARY. The maximum number of strong subtournaments with at least three nodes in any tournament T_n is

$$\begin{cases} 2^n - n2^{(1/2)(n-1)} - 1 & \text{if } n \text{ is odd,} \\ 2^n - 3n2^{(1/2)(n-4)} - 1 & \text{if } n \text{ is even.} \end{cases}$$

Let $s(n, k)$ denote the minimum number of strong subtournaments T_k that a strong tournament T_n can have. (If T_n is not strong, then it need not have any nontrivial strong subtournaments.) Moon (1966) discovered the following result.

THEOREM 5.   If $3 \leq k \leq n$, then $s(n, k) = n - k + 1$.

Proof. We first show that $s(n, k) \geq n - k + 1$. This inequality certainly holds when $n = k$ for any fixed value of k. It follows from Theorem 3 that any strong tournament T_n has a strong subtournament T_{n-1}. Now T_{n-1} has at least $s(n - 1, k)$ strong subtournaments T_k by definition and the node not in T_{n-1} is in at least one k-cycle. This k-cycle determines a strong subtournament T_k not contained in T_{n-1}. Consequently,

$$s(n, k) \geq s(n - 1, k) + 1.$$

The earlier inequality now follows by induction on n.

To show that $s(n, k) \leq n - k + 1$, consider the tournament T_n in which $p_i \rightarrow p_j$ when $i = j - 1$ or $i \geq j + 2$. (This tournament is illustrated in Figure 3.) It is easy to see that this tournament has precisely $n - k + 1$ strong subtournaments T_k if $3 \leq k \leq n$. This completes the proof of the theorem.

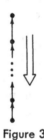

Figure 3

Notice that Theorem 5 remains true if the phrase "strong subtournaments T_k" is replaced by "k-cycles" in the definition of $s(n, k)$. The case $k = 3$ of this result was given by Harary, Norman, and Cartwright (1965, p. 306).

COROLLARY.   The minimum number of cycles a strong tournament T_n can have is $\binom{n-1}{2}$.

This is proved by summing the expression for $s(n, k)$ from $k = 3$ to $k = n$.

Exercises

1. Show that a sum of the type $\Sigma_{i=1}^{n} \binom{s_i}{t}$, where t is a fixed integer and $(s_1, s_2, \cdots, s_n)$ is the score vector of a tournament T_n, attains its minimum when the scores are as nearly equal as possible.

2. Prove that Property (b) implies Property (a).

3. Prove that, if a tournament has any cycles, then it has some 3-cycles.

4. Suppose the tournament T_n has at least one cycle; if every subtournament of T_n either is strong or has no cycles, then is T_n necessarily regular?

5. Show that

$$c_3(T_n) = \frac{n(n^2 - 1)}{24} - \frac{1}{2} \sum_{i=1}^{n} \left(s_i - \frac{n-1}{2} \right)^2.$$

6. Prove that T_n must be strong if it has more 3-cycles than a tournament T_{n-1} can have. [L. W. Beineke.]

7. Use Theorem 4 to determine for what values of n and k there exist tournaments T_n with property P_k. (See Exercise 4.2.)

8. An m by n *bipartite tournament* consists of two sets of nodes P and Q, where $|P| = m$ and $|Q| = n$, such that each node of P is joined with each node of Q by an arc; there are no arcs joining nodes in the same node-set. The score vectors of a bipartite tournament are defined in the same way as the score vector of an ordinary tournament. Give an example of two bipartite tournaments with the same score vectors that do not have the same number of 4-cycles.

9. Show that the maximum number of 4-cycles an m by n bipartite tournament can have is $[m^2/4] \cdot [n^2/4]$. [Moon and Moser (1962a).]

6. The Distribution of 3-cycles in a Tournament

In applying the method of paired comparisions, it is usually assumed that the objects can be ordered on a linear scale. Hence, the presence of cycles in the tournament representing the outcomes of the comparisons indicates inconsistency on the part of the judge or that the underlying assumption is inappropriate. Kendall and Babington Smith (1940) proposed the number of 3-cycles (suitable normalized, so as to equal 1 when there are no 3-cycles and 0 when there are as many 3-cycles as possible) as a measure of the consistency of the comparisons. The number of 3-cycles was chosen, because it is a simple function of the score vector of the tournament and hence is easy to calculate. To define significance tests for the departure from consistency, it is necessary to know the distribution of the number of 3-cycles in a random tournament. Tables of the distribution of the number of 3-cycles in a random tournament T_n have been given by Kendall and Babington Smith when $3 \leq n \leq 7$ and by Alway (1962b) when $8 \leq n \leq 10$. Kendall and Babington Smith conjectured and Moran (1947) proved that the distribution of the number of 3-cycles is asymptotically normal.

THEOREM 6.   Let c_n denote the number of 3-cycles in a random tournament T_n. Then the distribution of $(c_n - \mu')/\sigma$ tends to the normal distribution with zero mean and unit variance, where

$$\mu' = \frac{1}{4} \binom{n}{3} \quad \text{and} \quad \sigma^2 = \frac{3}{16} \binom{n}{3}.$$

Proof. In a tournament T_n, let the variable $t(ijk)$ be 1 or 0 according as the distinct nodes p_i, p_j, and p_k do or do not determine a 3-cycle. Since only two of the eight equally likely ways of orienting the arcs joining these nodes yield a 3-cycle, it follows that $E[t(ijk)] = \frac{1}{4}$. Therefore,

$$\mu' = E(c_n) = \frac{1}{4} \binom{n}{3}.$$

If $r(ijk) = t(ijk) - \frac{1}{4}$, then

$$\sigma^2 = E[(c_n - \mu')^2] = E[(\sum r(ijk))^2],$$

where the sum is over the triples i, j, and k. The products in this expansion are of the following types:

$$r(ijk) \cdot r(ijk), \quad r(ijk) \cdot r(ijw), \quad r(ijk) \cdot r(ivw), \quad \text{and} \quad r(ijk) \cdot r(uvw).$$

The variables appearing in each of the last two products are independent; hence, the expectation of their product equals the product of their individual expectations, zero. The variables in the second product are also independent, since the probability that three nodes form a 3-cycle is still $\frac{1}{4}$, even when the orientation of one of the arcs joining them is specified. Thus the only nonzero contributions to σ^2 come from products of the first type. Therefore,

$$\sigma^2 = \binom{n}{3} E[r(ijk)^2] = \frac{3}{16} \binom{n}{3}.$$

We shall now show that for each fixed positive integer h

(i) $\dfrac{\mu_{2h}}{\sigma^{2h}} \to \dfrac{(2h)!}{2^h h!}$ and   (ii) $\dfrac{\mu_{2h+1}}{\sigma^{2h+1}} \to 0$

as n tends to infinity (μ_k denotes the kth central moment of c_n). It will then follow from the second limit theorem of probability theory [see Kendall and Stuart (1958, p. 115)] that the distribution of $(c_n - \mu')/\sigma$ tends to the normal distribution with zero mean and unit variance.

If we combine all terms in the expansion of $\mu_{2h} = E[(\Sigma\, r(ijk))^{2h}]$ in which similar combinations of values of i, j, and k occur, as we did earlier for the second moment, the number of times terms of a given type appear will be a polynomial in n whose degree equals the number of different values of i, j, and k that occur in the terms. Therefore, μ_{2h} is a polynomial in n whose coefficients depend only on h. To calculate its degree and leading coefficient, we must identify the terms with a nonzero expectation in the expansion that involve the largest number of different values of the indices i, j, and k.

We may suppose that each term in the expansion has been split into classes of products such that different classes of the same term have no indices in common; we may also suppose that any factor in a class of more than one factor has at least one index in common with at least one other factor in the same class. If a term contains a factor r involving one or more indices that occur nowhere else in the term, then the term has expectation zero, since r is independent of the remaining factors in the term. Hence, if a term is to have a nonzero expectation, each index in it must occur at least twice.

We need only consider terms in which each index occurs exactly twice, because, as we shall show, the leading term in μ_{2h} is of order $3h$. Suppose such a term contains fewer than h classes. Some class of this term must contain at least three different factors. For any factor $r(ijk)$ of this class, either the indices i, j, and k occur in three other factors or one index, i say, occurs in another factor and j and k occur together in another. In either case $r(ijk)$ is independent of the remaining factors and the expectation of the term is zero. Consequently, the terms with a nonzero expectation that involve the largest number of different values of i, j, and k are of the type $\prod_{v=1}^{h} r(i_v j_v k_v)^2$. There are

$$\frac{(2h)!}{2^h h!} \binom{n}{3}\binom{n-3}{3}\cdots\binom{n-3(h-1)}{3}$$

terms of this type and the expected value of each is $(3/16)^h$. Therefore,

$$\mu_{2h} = \frac{(2h)!}{2^{6h} h!} n^{3h} + O(n^{3h-1}).$$

Since

$$\sigma^{2h} = \frac{n^{3h}}{2^{5h}} + O(n^{3h-1}),$$

it follows that (i) holds for each fixed positive integer h as n tends to infinity.

Similarly, if a term in the expansion of μ_{2h+1} is to have a nonzero expectation, each index in it must occur at least twice. Hence, μ_{2h+1} is a polynomial in n of degree at most $[3(2h+1)/2] = 3h+1$. Since σ^{2h+1} is of degree $3h + 3/2$ in n, it follows that (ii) also holds for each fixed positive integer h as n tends to infinity. This completes the proof of the theorem.

Notice that the expected value of c_n is asymptotically equal to the maximum value of c_n (see the first corollary to Theorem 4). This phenomenom occurs frequently in problems of this type.

The following theorem may be proved in the same way.

THEOREM 7.   Let q_n denote the number of 4-cycles in a random tournament T_n. Then the distribution of $(q_n - \mu')/\sigma$ tends to the normal distribution with zero mean and unit variance, where

$$\mu' = \frac{3}{8}\binom{n}{4} \quad \text{and} \quad \sigma^2 = \frac{3}{64}\binom{n}{4}(4n - 11).$$

Moon and Moser (1962a) proved the following analogous result for bipartite tournaments.

THEOREM 8.   Let $c(m, n)$ denote the number of 4-cycles in a random m by n bipartite tournament. Then, subject to certain mild conditions on the relative rates of growth of m and n, the distribution of $(c(m, n) - \mu')/\sigma$ tends to the normal distribution with zero mean and unit variance, where

$$\mu' = \frac{1}{8}\binom{m}{2}\binom{n}{2} \quad \text{and} \quad \sigma^2 = \frac{1}{64}\binom{m}{2}\binom{n}{2}(2m + 2n - 1).$$

Exercises

1. The quantity

$$h = \frac{12}{n^3 - n} \cdot \sum_{i=1}^{n} \left(s_i - \frac{1}{2}(n - 1) \right)^2$$

is called the *hierarchy index* of a tournament. Show that the mean and variance of h satisfy the equations $\mu' = 3/(n + 1)$ and $\sigma^2 = 18(n - 2)/(n + 1)^2 n(n - 1)$. [Landau (1951 a,b).]

2. Prove Theorem 7.

3. Prove that the third and fourth moments of the distribution of c_n are given by the formulas:

$$\mu_3 = -\frac{3}{32}\binom{n}{3}(n - 4)$$

and

$$\mu_4 = \frac{3}{256}\binom{n}{3}\left\{ 9\binom{n-3}{3} + 39\binom{n-3}{2} + 9\binom{n-3}{1} + 7 \right\}.$$

[Moran (1947).]

4. Let $q = q(n, k)$ denote the number of k-cycles in a random tournament $T_n(k \geq 3)$. Show that the mean and variance of q satisfy the equations

$$\mu' = \binom{n}{k}\frac{(k - 1)!}{2^k} \quad \text{and} \quad \sigma^2 = 0(n^{2k-3})$$

for each fixed value of k as n tends to infinity. What can be deduced about the number of k-cycles in most tournaments T_n?

5. Obtain a similar result for paths of length k.

7.   Transitive Tournaments

A tournament is *transitive* if, whenever $p \to q$ and $q \to r$, then $p \to r$ also. Transitive tournaments have a very simple structure. The following theorem gives some properties of a transitive tournament T_n whose scores $(s_1, s_2, \cdots, s_n)$ are in nondecreasing order.

THEOREM 9. The following statements are equivalent.

(1) T_n is transitive.

(2) Node p_j dominates node p_i if and only if $j > i$.

(3) T_n has score vector $(0, 1, \cdots, n - 1)$.

(4) The score vector of T_n satisfies the equation

$$\sum_{i=1}^{n} s_i^2 = \frac{n(n-1)(2n-1)}{6}.$$

(5) T_n contains no cycles.

(6) T_n contains exactly $\binom{n}{k+1}$ paths of length k, if $1 \leq k \leq n - 1$.

(7) T_n contains exactly $\binom{n}{k}$ transitive subtournaments T_k, if $1 \leq k \leq n$.

(8) Each principal submatrix of the dominance matrix $M(T_n)$ contains a row and column of zeros.

Every tournament T_n ($n \geq 4$) contains at least one transitive subtournament T_3, but not every tournament T_n is itself transitive. The following question arises: What is the largest integer $v = v(n)$ such that every tournament T_n contains a transitive subtournament T_v? Erdös and Moser (1964) gave the following bounds for $v(n)$. [The lower bound was first found by Stearns (unpublished).]

THEOREM 10. $[\log_2 n] + 1 \leq v(n) \leq [2 \log_2 n] + 1$.

Proof. Consider a tournament T_n in which the node p_n has the largest score s_n. It must be that $s_n \geq [\frac{1}{2}n]$, so there certainly exists a subtournament $T_{[(1/2)n]}$ in T_n each node of which is dominated by p_n. We may suppose that this subtournament contains a transitive subtournament with at least $[\log_2 [\frac{1}{2}n]] + 1$ nodes. These nodes together with p_n determine a transitive subtournament of T_n with at least

$$[\log_2 [\frac{1}{2}n]] + 2 = [\log_2 n] + 1$$

nodes. The lower bound now follows by induction.

There are $2^{\binom{n}{2} - \binom{v}{2}}$ tournaments T_n, containing a given transitive subtournament T_v, and there are $\binom{n}{v} v!$ such subtournaments T_v possible. Therefore,

$$\binom{n}{v} v! \, 2^{\binom{n}{2} - \binom{v}{2}} \geq 2^{\binom{n}{2}},$$

since every tournament T_n contains at least one transitive subtournament

T_v. This inequality implies that $n^v \geq 2^{\binom{v}{2}}$. Consequently, $v \leq [2 \log_2 n] + 1$, and theorem is proved.

The exact value of $v(n)$ is known only for some small values of n. For example, $v(7) \geq 3$ by Theorem 10. The tournament T_7 in which $p_i \to p_j$ if and only if $j - i$ is a quadratic residue modulo 7 contains no transitive subtournament T_4. It follows that $v(7) = 3$. Bent (1964) examined other similarly constructed tournaments and deduced the information about $v(n)$ given in Table 2. Erdös and Moser conjecture that $v(n) = [\log_2 n] + 1$ for all n.

Table 2. **$v(n)$, the largest integer v such that every tournament T_n contains a transitive subtournament T_v.**

$$v(2) = v(3) = 2$$
$$v(4) = \cdots = v(7) = 3$$
$$v(8) = \cdots = v(11) = 4$$
$$4 \leq v(12) \leq \cdots \leq v(15) \leq 5$$
$$v(16) = \cdots = v(23) = 5$$
$$5 \leq v(24) \leq \cdots \leq v(31) \leq 7$$
$$6 \leq v(32) \leq \cdots \leq v(43) \leq 7$$

Moon (1966) found the next two theorems.

Let $u(n, k)$ denote the maximum number of transitive subtournaments T_k that a strong tournament T_n can have. (The problem is trivial if T_n is not strong.)

THEOREM 11.   If $3 \leq k \leq n$, then $u(n, k) = \binom{n}{k} - \binom{n-2}{k-2}$.

Proof. When $k = 3$ the theorem follows from Theorem 5, since every subtournament T_3 is either strong or transitive. We now show that $u(n, k) \leq \binom{n}{k} - \binom{n-2}{k-2}$ for any larger fixed value of k. This inequality certainly holds when $n = k$. If $n > k \geq 4$, then any strong tournament T_n contains a strong subtournament T_{n-1} by Theorem 3. Let p be the node not in T_{n-1}; there are at most $u(n - 1, k - 1)$ transitive subtournaments T_k of T_n that contain the node p and at most $u(n - 1, k)$ that do not. We may suppose

$$u(n - 1, k - 1) \leq \binom{n-1}{k-1} - \binom{n-3}{k-3}$$

and

$$u(n - 1, k) \leq \binom{n-1}{k} - \binom{n-3}{k-2}.$$

Therefore,

$$u(n, k) \leq u(n - 1, k - 1) + u(n - 1, k)$$

$$\leq \binom{n-1}{k-1} + \binom{n-1}{k} - \binom{n-3}{k-3} - \binom{n-3}{k-2}$$

$$= \binom{n}{k} - \binom{n-2}{k-2}.$$

The inequality now follows by induction.

To show that $u(n, k) \geq \binom{n}{k} - \binom{n-2}{k-2}$ consider the strong tournament T_n in which $p_1 \to p_n$ but otherwise $p_j \to p_i$ if $j > i$. (This tournament is illustrated in Figure 4.) This tournament has exactly $\binom{n}{k} - \binom{n-2}{k-2}$ transitive subtournaments T_k if $3 \leq k \leq n$, because every subtournament T_k is transitive except those containing both p_1 and p_n. This completes the proof of the theorem.

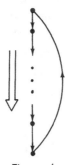

Figure 4

COROLLARY.   The maximum number of transitive subtournaments a strong tournament T_n ($n \geq 3$) can contain, including the trivial tournaments T_1 and T_2, is $3 \cdot 2^{n-2}$.

Let $r(n, k)$ denote the minimum number of transitive subtournaments T_k a tournament T_n can have. It follows from Theorem 10 that $r(n, k) = 0$ if $k > [2 \log_2 n] + 1$ and that $r(n, k) > 0$ if $k \leq [\log_2 n] + 1$.

THEOREM 12.   Let

$$\tau(n, k) = \begin{cases} n \cdot \dfrac{(n - 1)}{2} \cdot \dfrac{(n - 3)}{4} \cdots \dfrac{(n - 2^{k-1} + 1)}{2^{k-1}} & \text{if } n > 2^{k-1} - 1, \\ 0 & \text{if } n \leq 2^{k-1} - 1. \end{cases}$$

Then

$$r(n, k) \geq \tau(n, k).$$

Proof. When $k = 1$, the result is certainly true if we count the trivial tournament T_1 as transitive. If $k \geq 2$, then clearly

$$r(n, k) \geq \sum_{i=1}^{n} r(s_i, k - 1),$$

where $(s_1, s_2, \cdots, s_n)$ denotes the score vector of the tournament T_n. We may suppose that $r(s_i, k - 1) \geq \tau(s_i, k - 1)$; since $\tau(n, k)$ is a convex function of n for fixed values of k, we may apply Jensen's inequality and conclude that

$$r(n, k) \geq \sum_{i=1}^{n} \tau(s_i, k - 1) \geq n\tau \left(\tfrac{1}{2}(n - 1), k - 1\right) = \tau(n, k).$$

The theorem now follows by induction on k.

Notice that the lower bound in Theorem 10 follows from Theorem 12.

Exercises

1. Prove Theorem 9. [The equivalence between (4) and (5) is due to Moser; see Bush (1961). In connection with (6), see (Szele (1943) and Camion (1959); in connection with (8), see Marimont (1959) and Harary (1960).]

2. Construct the tournament T_7 defined after the proof of Theorem 10 and verify that it contains no transitive subtournament T_4.

3. Prove the corollary to Theorem 11.

4. Use Theorem 10 to prove that

$$r(n, k) \geq \binom{n}{2^{k-1}} \Big/ \binom{n-k}{2^{k-1}-k} \geq \left(\frac{n}{2^{k-1}}\right)^k \qquad \text{if } n \geq 2^{k-1}.$$

How does this bound compare with the one in Theorem 12?

5. Prove that $r(n, 3) = \tau(n, 3)$ when n is odd.

6. What is the maximum number of arcs $\overrightarrow{pq}$ a tournament T_n can have such that $\overrightarrow{pq}$ is not contained in any transitive subtournament T_3?

7. A transitive subtournament of T_n is said to be *maximal* if it is not contained in any larger transitive subtournament of T_n. Let $m(n)$ denote the maximum number of maximal transitive subtournaments that a tournament T_n can have. Show that $m(n) \geq (7)^{n/5} > (1.47)^n$ if $n \equiv 0(\bmod 5)$. Can this lower bound be improved? Try to obtain a nontrivial upper bound for $m(n)$.

8. If k is any fixed integer greater than two, let $t = t(n, k)$ denote the number of transitive subtournaments T_k contained in a random tournament T_n.

Show that the distribution of $(t - \mu')/\sigma$ tends to the normal distribution with zero mean and unit variance as n tends to infinity, where

$$\mu' = (n)_k 2^{-\binom{k}{2}}$$

and

$$\sigma^2 = \left(k! 2^{-\binom{k}{2}}\right)^2 \sum_{r=3}^{k} \binom{n}{k}\binom{k}{r}\binom{n-k}{k-r}\left(\frac{2^{\binom{r}{2}}}{r!} - 1\right).$$

8.   Sets of Consistent Arcs in a Tournament

In the last section, we considered subsets of nodes of a tournament such that the subtournament determined by these nodes was transitive. We may also consider subsets of arcs of a tournament T_n such that these arcs, by themselves, define no intransitivities. More specifically, we shall call the arcs in a set S *consistent* if it is possible to relabel the nodes of T_n in such a way that, if the arc $\overrightarrow{p_j p_i}$ is in S, then $j > i$. (An equivalent definition is that T_n contains no cycles all of whose arcs belong to S.) Erdös and Moon (1965) gave the following result.

THEOREM 13.   Let $w(n)$ denote the largest integer w such that every tournament T_n contains a set of w consistent arcs. Then

$$w(n) \geq \left[\frac{n}{2}\right] \cdot \left[\frac{n+1}{2}\right] \qquad \text{for all } n \text{ and } w(n) \leq \frac{(1+\epsilon)}{2}\binom{n}{2}$$

for any positive ϵ and all sufficiently large n.

Proof. When $n = 1$, the lower bound is trivially true. In any tournament T_n, there exists at least one node, say p_n, whose score is at least $\left[\frac{n}{2}\right]$. We may suppose that the tournament defined by the remaining $n - 1$ nodes contains a set S of at least $\left[\frac{n-1}{2}\right] \cdot \left[\frac{n}{2}\right]$ consistent arcs. The arcs in S and the arcs oriented away from P_n are clearly consistent. Therefore, T_n contains a set of at least

$$\left[\frac{n}{2}\right] + \left[\frac{n-1}{2}\right] \cdot \left[\frac{n}{2}\right] = \left[\frac{n}{2}\right] \cdot \left[\frac{n+1}{2}\right]$$

consistent arcs. The lower bound follows by induction.

We now prove the upper bound. Let ϵ be chosen satisfying the inequality $0 < \epsilon < 1$. The tournament T_n has $N = \binom{n}{2}$ pairs of distinct nodes and the nodes can be labeled in $n!$ ways. Hence, there are at most $n! \cdot \binom{N}{k}$ tournaments T_n, whose largest set of consistent arcs contains k arcs. So, an upper bound for the number of tournaments T_n that contain a set of more than $\dfrac{(1+\epsilon)}{2} N$ consistent arcs is given by

$$n! \cdot \sum_{k > (1+\epsilon)N/2} \binom{N}{k} < n! \, N \binom{N}{[(1+\epsilon)N/2]} \le n! \, N 2^N \binom{N}{[(1+\epsilon)N/2]} \cdot \binom{N}{[\frac{1}{2}N]}^{-1}$$

$$= n! \, N 2^N \frac{\left(N - [\frac{1}{2}N]\right)_{[(1+\epsilon)N/2] - [\frac{1}{2}N]}}{\left([(1+\epsilon)N/2]\right)_{[(1+\epsilon)N/2] - [\frac{1}{2}N]}}$$

$$\le n! \, N 2^N \left(\frac{N - [\frac{1}{2}N]}{[(1+\epsilon)N/2]}\right)^{[\epsilon N/2]}$$

$$\le n! \, N 2^N \left(1 - \frac{1}{2}\epsilon\right)^{[\epsilon N/2]},$$

provided that n is sufficiently large. Since $1 - x < e^{-x}$ when $0 < x < 1$, it follows that this last expression is less than

$$n! \, N 2^N e^{-(\epsilon/2)[\epsilon N/2]};$$

but this in turn is less than 2^N for all sufficiently large values of n. Hence, there must be at least one tournament T_n that does not contain any set of more than $\dfrac{(1+\epsilon)}{2} N$ consistent arcs. This completes the proof of the theorem.

A more careful analysis of the above inequalities shows that the proportion of tournaments T_n that contain more than

$$\frac{1}{2}\binom{n}{2} + (1 + \epsilon)\left(\frac{1}{2} n^3 \log n\right)^{1/2}$$

consistent arcs, for any fixed positive ϵ, tends to zero as n tends to infinity.

Exercises

1. Determine the exact value of $w(n)$ for $3 \le n \le 5$.

2. Let $z = z(n)$ denote the least integer such that any two tournaments with n nodes can be made isomorphic by reversing at most z arcs of one of the tournaments. Prove that $z(n) = \left(\frac{1}{2} + o(1)\right)\binom{n}{2}$.

3. Prove the assertion in the last paragraph of this section.

4. An *oriented graph*, or incomplete tournament, $T(n, m)$ consists of n nodes, m pairs of which are joined by a single arc. Let $w = w(n, m)$ denote the largest integer such that every oriented graph $T(n, m)$ contains a set of w consistent arcs. Prove that $\lim_{n \to \infty} w(n, m)/m = 1/2$, under suitable assumptions on the relative rates of growth of m and n. [Erdös and Moon (1965).]

5. Let $u(T_n)$ denote the least number of arcs possible in a subset U of the arcs of T_n if at least one arc of every cycle of T_n belongs to U. Let $r(T_n)$ denote the smallest integer r such that T_n can be transformed into a transitive tournament by reversing the orientation of r arcs. Finally, let $w(T_n)$ denote the maximum number of consistent arcs in T_n. Prove that $u(T_n) = r(T_n) = \binom{n}{2} - w(T_n)$ for any tournament T_n. [Remage and Thompson (1966) gave an algorithm for determining $r(T_n)$.]

6. Let $f(n, i)$ denote the number of tournaments T_n such that $r(T_n) = i$. Prove the following equations [Slater (1961)].

$$f(n, 0) = n!,$$

$$f(n, 1) = \frac{n!(3n^2 - 13n + 14)}{6},$$

$$f(n, 2) = \frac{n!(9n^4 - 78n^3 + 235n^2 - 438n + 680)}{72} \qquad \text{if } n \geq 4,$$

$$f(n, 3) =$$
$$\begin{cases} 24 \quad \text{if } n = 5, \\ \dfrac{n!(135n^6 - 1755n^5 + 8685n^4 - 27{,}185n^3 + 77{,}820n^2 - 157{,}204n + 210{,}336)}{6480} \\ \qquad\qquad\qquad\qquad\qquad\qquad\qquad\qquad\qquad\qquad\qquad \text{if } n \geq 6. \end{cases}$$

9. The Parity of the Number of Spanning Paths of a Tournament

Every tournament has a spanning path. This is an immediate consequence of Theorem 3, if the tournament T_n is strong. If T_n is not strong, then it has a unique decomposition into strong subtournaments $T^{(1)}, T^{(2)}, \cdots, T^{(l)}$ such that every node of $T^{(j)}$ dominates every node of $T^{(i)}$ if $1 \leq i < j \leq l$. Each of these strong subtournaments has a spanning path and these spanning paths can be combined to yield a path spanning T_n.

The following theorem is a special case of a result proved by Rédei (1934) and generalized by Szele (1943).

THEOREM 14. Every tournament has an odd number of spanning paths.

Proof. Let $A = [a_{ij}]$ denote the dominance matrix of an arbitrary tournament T_n. Recall that $a_{ij} = 1$ if $p_i \rightarrow p_j$ and $a_{ij} = 0$ otherwise; in particular, $a_{ii} = 0$. The number h of spanning paths in T_n is equal to the sum of all products of the type

$$a_{i_1\pi(i_1)}a_{i_2\pi(i_2)} \cdots a_{i_{n-1}\pi(i_{n-1})},$$

where $1 \leq i_1 < i_2 \cdots < i_{n-1} \leq n$ and π is a permutation of the set $N = \{1, 2, \cdots, n\}$ such that $\pi^k(i) \neq i$ for $1 \leq k \leq n - 1$.

We introduce more notation before continuing the proof. The symbols $e, e_1, \cdots$ denote subsets of the set N; e' is the complement of e with respect to N; $e(M)$ is the determinant of the submatrix of the n by n matrix M whose row and column indices belong to e; if e is the empty set, then $e(M) = 1$. The determinant of M will be denoted by $|M|$.

Let A_k denote the matrix obtained from A by replacing the kth column by a column of 1's. Set

$$S_k = \sum_{1 \in e} e(A_k) \cdot e'(A_k),$$

where the sum is over all subsets e of N that contain 1. (The restriction that $1 \in e$ merely serves to distinguish between e and e'.) We now show that

$$h \equiv \sum_{k=1}^{m} S_k \quad \text{(mod 2)}. \tag{1}$$

It is clear that any term g in the expansion of S_k corresponds to a term in the expansion of the determinant of A_k. (Since the summation is modulo 2, we may disregard the signs of the terms.) Suppose the permutation of N associated with this term can be expressed as the product of the disjoint cycles $c_1, c_2, \cdots, c_r$. Those nonzero terms associated with permutations for which $r = 1$, arising only when $e = N$, are precisely those terms described earlier whose sum is h. To prove assertion (1), it will suffice to show that the sum of the other terms is even.

Consider any term g for which $r > 1$. There is no loss of generality if we assume that $1 \in c_1$. Now, all the elements in c_i belong to e or they all belong to e', where e is a subset of N giving rise to g in the expansion of S_k. Since there are two choices for each cycle $c_2, c_3, \cdots, c_r$ it follows that there are 2^{r-1} subsets e giving rise to g. Consequently, the term g appears 2^{r-1} times in the sum. This completes the proof of (1), since 2^{r-1} is even if $r > 1$.

Let $\bar{A} = [\bar{a}_{ij}]$ denote the matrix whose elements satisfy the equation $\bar{a}_{ij} = a_{ij} + 1$ for all i and j. This implies that

$$|\bar{A}| = |A| + \sum_{k=1}^{n} |A_k|. \tag{2}$$

We observe that $\bar{a}_{ij} \equiv a_{ji} \pmod{2}$ if $i \neq j$ and that $\bar{a}_{ii} = 1$. Hence, the transpose of $\bar{A}$, which has the same determinant as $\bar{A}$, has its entries all congruent to the corresponding entries of A with the exception of the diagonal entries. Consequently,

$$\sum_{k=1}^{n} |A_k| = |\bar{A}^T| - |A| \equiv \sum_{e \neq N} e(A) \quad (\text{mod } 2), \quad (3)$$

since the last sum effectively involves those terms in the expansion of $|\bar{A}^T|$ that contain diagonal entries. (Notice that the diagonal entries in a term in the expansion of $|A^T|$ corresponding to some term in the expansion of $e(A)$ are those $\bar{a}_{ii}$ for which $i \in e'$; this is why the sum is over the proper subsets e of N.)

Let $\Sigma_k^* e(A_k)$ denote the sum of $e(A_k)$ over all k such that $k \in e$ with the convention that an empty sum equals zero. Then,

$$h \equiv \sum_{k=1}^{n} \left\{ \sum_{1 \in e} e(A_k) \cdot e'(A_k) \right\}$$

$$\equiv \sum_{1 \in e} \left\{ \sum_{k \in e} e(A_k) \, e'(A_k) + \sum_{k \notin e} e(A_k) \cdot e'(A_k) \right\}$$

$$\equiv \sum_{1 \in e} \left\{ e'(A) \sum_k{}^* e(A_k) + e(A) \sum_k{}^* e'(A_k) \right\} \quad (4)$$

$$\equiv \sum_e e'(A) \sum_k{}^* e(A_k) \quad (\text{mod } 2).$$

Let us apply (3) to $e(A)$ instead of to $|A|$; then

$$\sum_k{}^* e(A_k) \equiv \sum_{e_1 \subset e}' e_1(A) \quad (\text{mod } 2), \quad (5)$$

where the second summation is over all proper subsets e_1 of e. (If e is the empty set, then both sides equal zero.) If we substitute (5) into (4), we find that

$$h \equiv \sum_e \sum_{e_1 \subset e} e'(A) \cdot e_1(A) \quad (\text{mod } 2). \quad (6)$$

If $e - e_1 \neq N$, then the term arising from e and e_1, where $e_1 \subset e$ and $e_1 \neq e$, is equal to the term arising from e_1' and e'. Hence, the terms in the sum come in pairs except for the terms arising when $e - e_1 = N$, that is, when $e = N$ and e_1 is the empty set. Since $e'(A) = e_1(A) = 1$ in this case, it follows that

$$h \equiv 1 \quad (\text{mod } 2).$$

This completes the proof of the theorem.

COROLLARY. A tournament T_n has at least $\binom{n}{k}$ paths of length $k - 1$ if $2 \leq k \leq n$; furthermore, the number of paths of length $k - 1$ in T_n is congruent to $\binom{n}{k}$ modulo 2.

This is proved by applying Theorem 14 to each subtournament T_k of T_n and then summing over all such subtournaments.

Let G denote a graph that differs from a tournament T_n in that some of the arcs $\overrightarrow{p_i p_j}$ of T_n may have been either removed entirely or replaced by unoriented edges $p_i p_j$. A path in such a graph G may pass through any unoriented edge $p_i p_j$ in either direction. Let $(G)_k$ denote the number of ways of labeling the nodes of G as $q_1, q_2, \cdots, q_n$ in such a way that there are k pairs of consecutive nodes q_i and q_{i+1} for which the arc $\overrightarrow{q_{i+1} q_i}$ is in G (notice that $(G)_0 + (G)_1 + \cdots + (G)_{n-1} = n!$), and let (G) denote the number of spanning paths of G. Finally, let $[G]_k$ denote the number of sets of $n - k$ disjoint paths of G with the property that every node of G belongs to one path in the set. (Notice that there are k arcs or edges involved in each such set and that $[G]_{n-1} = (G)_0 = (G)$ if G is a tournament.)

Szele (1943) generalized Theorem 14 by showing that if G is a tournament T with n nodes, then

$$(T)_k \equiv \binom{n-1}{k} \qquad (\text{mod } 2), \qquad (7)$$

and

$$[T]_k \equiv \left(\left[\frac{n+k-1}{2} \right] \atop k \right) \qquad (\text{mod } 2). \qquad (8)$$

Exercises

1. Give a direct proof of the result that every (finite) tournament has a spanning path.

2. Give an example of a tournament with an infinite number of nodes that does not have a spanning path. [See Korvin (1966).]

3. Prove Equation (2).

4. Let G denote the graph obtained from the tournament T_n by replacing an arbitrary arc $\overrightarrow{p_i p_j}$ by an unoriented edge $e = p_i p_j$ and let H denote the graph obtained from T_n simply by removing the arc $\overrightarrow{p_i p_j}$. Prove the following statements and use them to give another proof of Theorem 14.

(a) $(G) = \sum_{i=0}^{n-1} (-1)^i (n - i)! \, [H]_i$.

(b) $(G) \equiv (H) \pmod 2$.
(c) The number of spanning paths of G that contain the edge e is even.
(d) The parity of the number of spanning paths of T_n is not changed if the arc $\overrightarrow{p_i p_j}$ is replaced by the arc $\overrightarrow{p_j p_i}$. [Szele (1943).]

5. Prove Equation (7) by first showing that

$$(T)_k = \sum_{i=0}^{n-1-k} (-1)^i \binom{k+i}{k} (n - k - i)! \, [T]_{k+i}.$$

6. If T^* denotes a transitive tournament T_n, then show that $[T^*]_k$ is equal to the number of ways of partitioning n different objects into $n - k$ indistinguishable nonempty subsets. Deduce from this that

$$[T^*]_k = \frac{1}{(n-k)!} \sum_{i=0}^{n-k-1} (-1)^i \binom{n-k}{i}(n - k - i)^n = S(n, n - k),$$

where $S(m, t)$ denotes a Stirling number of the second kind. (The Stirling numbers $S(m, t)$ may be defined by the identity

$$x^m = \sum_{t=1}^{m} S(m, t)(x)_t,$$

where $(x)_t = x(x - 1) \cdots (x - t + 1)$; they satisfy the recurrence relation

$$S(m, t) = S(m - 1, t - 1) + tS(m - 1, t).)$$

7. Prove that

$$(T^*)_k = \sum_{t=0}^{n-k-1} (-1)^t \binom{n+1}{t}(n - k - t)^n.$$

8. Use Theorem 14 to prove that, if T is any tournament with n nodes, then

$$]T]_k \equiv [T^*]_k \qquad (\text{mod } 2).$$

9. Show that

$$[T^*]_k = S(n, n - k) = \sum j_1 j_2 \cdots j_k,$$

where the sum is over all sets of k integers j_i such that $1 \le j_1 \le j_2 \le \cdots \le j_k \le n - k$.

10. Use the results in Exercises 8 and 9 to prove Equation (8). [Hint: Which terms in the sum in Exercise 9 are odd? Recall that the number of ways of choosing t objects from m objects when repetitions are allowed is

$$\binom{m+t-1}{t}.]$$

11. Let k denote the maximum number of nodes that can be chosen from the oriented graph T such that no two of the chosen nodes are joined by an arc. Prove that there exists a set of t disjoint paths in T, where $t \le k$, such that every node of T belongs to exactly one path in the set. [Gallai and Milgram (1960).]

10.   The Maximum Number of Spanning Paths
of a Tournament

Let $t(n)$ denote the maximum number of spanning paths a tournament T_n can contain. The following result is due to Szele (1943).

THEOREM 15.   If

$$\alpha = \lim_{n \to \infty} \left(\frac{t(n)}{n!} \right)^{1/n},$$

then α exists and satisfies the inequality $.5 = 2^{-1}_{\bullet} \leq \alpha \leq 2^{-3/4} < .6$.

Proof. Let us assume for the time being that the limit does exist.

The expected number of spanning paths in a random tournament T_n is $\frac{n!}{2^{n-1}}$. Consequently,

$$t(n) \geq \frac{n!}{2^{n-1}} \qquad \text{or} \qquad \left(\frac{t(n)}{n!} \right)^{1/n} \geq \frac{1}{2^{1-1/n}} \tag{1}$$

for all positive integers n. This implies the first inequality of the theorem.

Let A, B, C, and D denote the number of subtournaments T_4 a given tournament T_n has of the four types depicted in Figure 1. Subtournaments of these types have 1, 3, 3, and 5 spanning paths, respectively. Hence, the total number of 3-paths (paths of length three) in the tournament T_n is equal to

$$A + 3(B + C) + 5D = \binom{n}{4} + 2(n - 3)\, c_3(T_n),$$

where $c_3(T_n)$ is the number of 3-cycles in T_n. (The equality is established by checking that each subtournament T_4 is counted the appropriate number of times by the right-hand expression.)

But the corollary to Theorem 4 gives an upper bound for $c_3(T_n)$. We find, therefore, that an upper bound for the number of 3-paths in any tournament T_n is given by

$$\begin{cases} \dfrac{n^2(n - 1)(n - 3)}{8} & \text{if } n \text{ is odd,} \\[3mm] \dfrac{(n + 1)\, n(n - 2)(n - 3)}{8} & \text{if } n \text{ is even.} \end{cases}$$

It follows that

$$t(n) \leq \begin{cases} \dfrac{n^2(n - 1)(n - 3)}{8} \cdot \dfrac{(n - 4)^2(n - 5)(n - 7)}{8} \cdots & \text{if } n \text{ is odd,} \\[3mm] \dfrac{(n + 1)\, n(n - 2)(n - 3)}{8} \cdot \dfrac{(n - 3)(n - 4)(n - 6)(n - 7)}{8} \cdots & \\[3mm] & \text{if } n \text{ is even.} \end{cases}$$

The last factors in these products will depend on the residue class modulo 4 to which n belongs. We find in all cases, however, that

$$t(n) \leq \frac{(n+1)!}{8^{[n/4]}} \leq \frac{(n+1)!}{2^{(3/4)n-3}}.$$

Therefore,

$$\left(\frac{t(n)}{n!}\right)^{1/n} < \frac{(n+1)^{1/n}}{2^{(3/4)-(3/n)}} \tag{2}$$

for all positive integers n. This implies the last inequality of the theorem.

We shall use the following result on subadditive functions [see Fekete (1923)] to show that the limit actually exists.

LEMMA. Let $\{g(n) : n = 1, 2, \cdots\}$ be a sequence such that $g(a + b) \leq g(a) + g(b)$ for all a, $b = 1, 2, \cdots$. If $\phi = \inf_n g(n)/n$ is finite, then $\lim_{n \to \infty} g(n)/n$ exists and equals ϕ.

Proof. Choose the integer r so that

$$\frac{g(r)}{r} \leq \phi + \epsilon,$$

for an arbitrary positive value of ϵ. For any given value of n, let $n = rs + t$, where $0 < t \leq r$. Then

$$\frac{g(n)}{n} = \frac{g(rs + t)}{n} \leq \frac{g(rs)}{n} + \frac{g(t)}{n}$$

$$\leq \frac{sg(r)}{n} + \frac{g(t)}{n} \leq \frac{rs}{n}(\phi + \epsilon) + \frac{g(t)}{n}$$

$$\leq \phi + \epsilon + \frac{1}{n} \max \{g(1), \cdots, g(r)\}.$$

Therefore,

$$\lim_{n \to \infty} \sup \frac{g(n)}{n} \leq \phi + \epsilon$$

for every positive ϵ. The lemma follows immediately.

In forming a spanning path of a tournament T_n, we may choose the first k nodes of this path, form a path spanning these k nodes, and then join this path to a path spanning the remaining $n - k$ nodes. Consequently,

$$t(n) \leq \binom{n}{k} t(k) \cdot t(n - k) \qquad \text{or} \qquad \frac{t(n)}{n!} \leq \frac{t(k)}{k!} \cdot \frac{t(n - k)}{(n - k)!}$$

for all positive integers k and n with $k < n$. If we let $h(n) = \frac{t(n)}{n!}$, then

$$\log h(a + b) \leq \log h(a) + \log h(b)$$

for all positive integers a and b. The existence of $\lim\limits_{n \to \infty} \left(\dfrac{t(n)}{n!} \right)^{1/n}$ may now be demonstrated by applying the lemma to the function $\log h(n)$.

Szele conjectured that the limit actually is $\frac{1}{2}$ and gave the following table of values of $t(n)$.

Table 3. $t(n)$, the maximum number of spanning paths in a tournament T_n.

n	3	4	5	6	7
$t(n)$	3	5	15	45	189

Exercises

1. Supply the details omitted in the proof of inequality (2).

2. The inequality $t(n) \leq n!(t(k)/k!)^{\lceil n/k \rceil}$ holds for any fixed value of k. How does this upper bound compare with (2) for small values of k?

3. Let $c(n, k)$ denote the maximum number of k-cycles possible in a tournament T_n. Prove that

$$\binom{n}{k} \frac{(k-1)!}{2^k} \leq c(n, k) \leq \binom{n}{k} \frac{(k+1)(k-1)!}{2^{(3/4)k-3}}.$$

[G. Korvin and others.]

4. The expected number of spanning cycles in a random tournament T_n is approximately $\dfrac{2\pi}{n} \left(\dfrac{n}{2e} \right)^n$. It seems difficult to give explicit examples of tournaments having this many spanning cycles in general. Construct a tournament having at least $\left(\dfrac{n}{3e} \right)^n$ spanning cycles for large values of n.

[L. Moser.]

11.   An Extremal Problem

A tournament T_n has *property* $S(k, m)$, where $k \leq n$, if for every subset A of k nodes there exist at least m nodes p such that p dominates every node of A. In 1962, K. Schütte raised the question of determining the least integer n such that there exist tournaments T_n with property $S(k, 1)$, if such tournaments exist at all. Erdös (1962) showed that such tournaments do indeed exist and that if the tournament T_n has property $S(k, 1)$ then

$$n \geq 2^{k+1} - 1, \qquad \text{for } k = 1, 2, \cdots. \tag{1}$$

E. and G. Szekeres (1965) deduced a stronger inequality as a consequence of the following result.

THEOREM 16. If the tournament T_n has property $S(k, m)$, then

$$n \geq 2^k(m + 1) - 1. \qquad (2)$$

Proof. If p is any node of the tournament T_n, let T_d denote the sub-tournament determined by the d nodes q that dominate p. We show that T_d has property $S(k - 1, m)$.

Suppose that $d \geq k - 1$. Let A be any set of $k - 1$ nodes of T_d; since T_n has property $S(k, m)$, there must exist a set B of at least m nodes q such that q dominates every node in the set $A \cup \{p\}$. Since all these nodes dominate p, it follows that they all are in the tournament T_d. Therefore T_d has property $S(k - 1, m)$.

The other alternative is that $d < k - 1$. In this case, adjoin to T_d any $k - 1 - d$ other nodes of T_n (but not p) to form a subtournament T_{k-1}. The same argument as used before shows that T_{k-1} has property $S(k - 1, m)$. But this is impossible, since it would imply that the nodes of T_{k-1} dominate themselves.

We now prove inequality (2) by induction; it certainly holds when $k = 0$ and $m \geq 1$, since we may say that a tournament T_n has property $S(0, m)$ if and only if $n \geq m$. Suppose therefore that $k > 0$ and that (2) has been proved for all tournaments with property $S(k - 1, m)$. It follows from the remarks in the preceding paragraphs and the induction hypothesis that at least $2^{k-1} (m + 1) - 1$ arcs are oriented toward each node of T_n. Therefore,

$$n(2^{k-1} (m + 1) - 1) \leq \binom{n}{2},$$

or

$$n \geq 2^k(m + 1) - 1.$$

COROLLARY. If the tournament T_n has property $S(k, 1)$, then

$$n \geq 2^{k-1} (k + 2) - 1, \qquad \text{for } k = 2, 3, \cdots.$$

Proof. If T_n has property $S(k, 1)$ where $k \geq 2$, then T_n has property $S(k - 1, k + 1)$. For, let A be any set of $k - 1$ nodes of T_n and suppose that the set B of nodes that dominate every node of A contains only h nodes, where $h \leq k$. Then there exists a node x that dominates every node in B. Furthermore, there exists a node y that dominates every node in $A \cup \{x\}$. Then y must be in B and x and y must dominate each other, an impossibility. The corollary now follows from Theorem 16.

This corollary is best possible when $k = 2, 3$ (see Exercises 1 and 2) but it is not known if it is best possible when $k > 3$.

Erdös (1962) showed that there exist tournaments T_n with property $S(k, 1)$ whenever

$$n > 2^k k^2 \log (2 + \epsilon) \qquad (3)$$

for any positive ϵ, provided that k is sufficiently large. The ideas used in establishing this and an analogous result for tournaments with property $S(k, m)$ are similar to those used in proving the following result due to Erdös and Moser (1964b).

If A and B are disjoint subsets of the tournament T_n, let $t = t(A, B)$ denote the number of nodes p that dominate every node of A and are dominated by every node of B. If ϵ is any positive number, let $F(n, k, \epsilon)$ denote the number of tournaments T_n such that

$$\left| t - \frac{n - k}{2^k} \right| < \frac{\epsilon(n - k)}{2^k} \tag{4}$$

for every pair of subsets A and B for which $|A| + |B| = k$.

THEOREM 17.   For every positive ϵ there exist constants c and K, which depend on ϵ, such that if $k > K$ and $n > ck^2 2^k$, then

$$\lim_{n \to \infty} \frac{F(n, k, \epsilon)}{2^{\binom{n}{2}}} = 1. \tag{5}$$

Proof. Suppose $0 < \epsilon < 1$. For a particular choice of A and B, there are

$$\binom{n-k}{t}(2^k - 1)^{n-k-t} \cdot 2^{\binom{n}{2} - k(n-k)} = 2^{\binom{n}{2}} \binom{n-k}{t} \left(\frac{1}{2^k}\right)^t \left(1 - \frac{1}{2^k}\right)^{n-k-t}$$

tournaments T_n for which $t(A, B) = t$. For, there are $\binom{n-k}{t}$ choices for the t nodes that dominate the nodes of A and are dominated by the nodes of B; the other $n - k - t$ nodes may each be joined to the k nodes of $A \cup B$ in one of $2^k - 1$ ways; the remaining $\binom{n}{2} - k(n - k)$ arcs may be oriented arbitrarily. There are $\binom{n}{2} 2^k$ ways of selecting the sets A and B. Consequently, if S denotes the number of tournaments T_n that do not satisfy (4) for every choice of A and B, then

$$S \leq \binom{n}{k} 2^{\binom{n}{2}+k} \sum' \binom{n-k}{t}\left(\frac{1}{2^k}\right)^t \left(1 - \frac{1}{2^k}\right)^{n-k-t},$$

where the sum is over all t such that $\left| t - \frac{(n - k)}{2^k} \right| \geq \frac{\epsilon(n - k)}{2^k}$.

The largest term in the sum occurs when $t = T = \left[(1 - \epsilon) \frac{(n-k)}{2^k} \right]$, $\binom{n}{k} 2^k \leq n^k$ if $k \geq 4$, and $1 - x < e^{-x}$ if $0 < x < 1$. Therefore,

$$S \leq 2^{\binom{n}{2}} n^{k+1} \binom{n-k}{T} \frac{1}{2^{kT}} e^{-(n-k-T)/2^k}$$

when $k \geq 4$. But

$$\binom{n-k}{T}\frac{1}{2^{kT}} < \left(\frac{(n-k)e}{T2^k}\right)^T \leq c_1 \left(\frac{e}{1-\epsilon}\right)^T,$$

where c_1 denotes some constant. Consequently, to prove (5), we need only show that

$$n^{k+1}\left(\frac{e}{1-\epsilon}\right)^{(n/2^k)(1-\epsilon)} e^{-(n/2^k)+(n/2^{2k})(1-\epsilon)}$$

tends to zero as n tends to infinity, since $\dfrac{k}{2^k}$ is bounded.

If L denotes the logarithm of this expression, then

$$L = (k+1)\log n - \frac{n}{2^k}\left[(1-\epsilon)\log(1-\epsilon) + \epsilon - \frac{(1-\epsilon)}{2^k}\right].$$

Since

$$\log(1-\epsilon) = -\epsilon - \tfrac{1}{2}\epsilon^2 - \tfrac{1}{3}\epsilon^3 - \cdots > -\epsilon - \tfrac{1}{2}\epsilon^2(1+\epsilon+\cdots)$$

$$= \frac{-\epsilon + \tfrac{1}{2}\epsilon^2}{1-\epsilon},$$

when $0 < \epsilon < 1$, it follows that, for sufficiently large values of k, the quantity inside the brackets will exceed some positive constant $c_2 = c_2(\epsilon)$. Therefore,

$$L \leq (k+1)\log n - \frac{c_2 n}{2^k} \leq n(k+1)\left(\frac{\log n}{n} - \frac{c_3}{k2^k}\right),$$

where c_3 is another constant equal, say, to $\tfrac{1}{2}c_2$. Now let us suppose that

$$n > ck^2 2^k,$$

where the constant $c = c(\epsilon)$ will be specified presently. Then it follows that

$$L \leq n(k+1)\left[\frac{\log(ck^2 2^k)}{ck^2 2^k} - \frac{c_3}{k2^k}\right] = \frac{n(k+1)}{ck2^k}\left[\frac{\log(ck^2)}{k} + \log 2 - cc_3\right],$$

since $\log n/n$ is a decreasing function for $n \geq 3$. It is clear that c can be defined as a function of ϵ in such a way that, if k is sufficiently large, then the expression within the brackets is negative. Consequently, L will tend to $-\infty$ as n tends to $+\infty$. This completes the proof of the theorem.

When this type of argument is applied to the simpler problem of establishing (3), it turns out that the constant c may be any quantity greater than $\log 2$.

Exercises

1. Let T_7 denote the tournament in which $p_i \to p_j$ if and only if $j - i$ is a quadratic residue modulo 7. Prove that T_7 has property $S(2, 1)$. [Erdös (1962).]

2. Let T_{19} denote the tournament in which $p_i \to p_j$ if and only if $j - i$ is a quadratic residue modulo 19. Prove that T_{19} has property $S(3, 1)$. [E. and G. Szekeres (1965).]

3. Prove that if a tournament T_n with property $S(k, 1)$ exists, then a tournament T_r with property $S(k, 1)$ exists whenever $r > n$. [See the remark at the end of the second paragraph of Erdös and Moser (1964b).]

4. Let A and B denote two disjoint subsets of nodes of a tournament T_n such that $|A| = a$ and $|B| = b$. If $t(A, B) \geq m$ for every choice of A and B, then we will say that T_n has property $T(a, b; m)$. (Notice that property $T(a, 0; m)$ is the same as property $S(a, m)$.) Generalize Theorem 16 by obtaining a lower bound for the number of nodes a tournament must have if it has property $T(a, b; m)$.

5. Let $h = h(n)$ denote the smallest integer such that in any tournament T_n there exists some set E of h nodes such that any node not in E is dominated by at least one node in E. Show that $[\log_2 n - 2 \log_2 \log_2 n] \leq h(n) \leq [\log_2 (n + 1)]$ if $n \geq 2$. [L. Moser.]

12. The Diameter of a Tournament

The *diameter* of a strong tournament T_n ($n \geq 3$) is the least integer d such that, for every ordered pair of nodes p and q of T_n, there exists a nontrivial path $P(p, q)$ of length at most d. (The diameter of a reducible tournament is not defined.) No node is connected to itself by a nontrivial path of length less than three so the diameter of every strong tournament is at least three.

We shall prove a result that implies that almost all tournaments T_n (that is, all but a fraction that tends to zero as n tends to infinity) have the following property: the number of paths of length two from p_i to p_j lies between $(\frac{1}{4} - \epsilon)(n - 2)$ and $(\frac{1}{4} + \epsilon)(n - 2)$ for every ordered pair of distinct nodes p_i and p_j and for any positive ϵ. Consequently, almost all tournaments have diameter three, since there is a path of length three connecting every node to itself in every strong tournament (and almost all tournaments are strong by Theorem 1).

Let r_{ij} denote the number of paths of length two from p_i to p_j ($p_i \neq p_j$) in a tournament T_n and let $E(n, \lambda)$ denote the expected number of ordered pairs of distinct nodes p_i and p_j in a random tournament T_n for which $|r_{ij} - \frac{1}{4}(n - 2)| > \lambda$. The following theorem is very similar to a theorem proved by Moon and Moser (1966).

THEOREM 18.   If $\lambda = \{\frac{3}{4}(n - 2)[\log (n - 2) + w(n)]\}^{1/2}$, where $\log (n - 2) + w(n) \to \infty$ and $w(n)n^{-1/3} \to 0$, then

$$E(n, \lambda) \sim \{2\pi[\log(n - 2) + w(n)]\}^{-1/2} e^{-2w(n)}$$

as $n \to \infty$.

Proof. There are

$$\binom{n-2}{r} 3^{n-2-r} 2^{\binom{n}{2}-2(n-2)} = 2^{\binom{n}{2}} \binom{n-2}{r}\left(\frac{1}{4}\right)^r\left(\frac{3}{4}\right)^{n-2-r}$$

tournaments T_n such that $r_{ij} = r$ for any admissible choice of i, j, and r. For, there are $\binom{n-2}{r}$ ways of selecting the r nodes that are to be on a path of length two from p_i to p_j and there are three ways of orienting the arcs joining p_i and p_j to any node that is not on such a path; the remaining $\binom{n}{2} - 2(n-2)$ arcs may be oriented arbitrarily. Since there are $n(n-1)$ ordered pairs of distinct nodes p_i and p_j, it follows that

$$E(n, \lambda) = n(n-1) \sum \binom{n-2}{r}\left(\frac{1}{4}\right)^r\left(\frac{3}{4}\right)^{n-2-r}$$

where the sum is over all r such that $|r - \frac{1}{4}(n-2)| > \lambda$. We may apply the De Moivre-Laplace Theorem to the sum [see Feller (1957; p. 172)] and conclude that

$$E(n, \lambda) \sim 2n(n-1)(1 - \Phi(X)),$$

where $X = 2(2\lambda + 1)[3(n-2)]^{-1/2}$ and Φ is the normal distribution function. (We need the hypothesis that $w(n)n^{-1/3} \to 0$ at this step.)

Since $X \to \infty$ as $n \to \infty$, we may use the relation

$$1 - \Phi(X) \sim \frac{1}{(2\pi)^{1/2}X} e^{-(1/2)X^2}$$

[see Feller (1957; p. 166)]. After further simplification, we find that

$$E(n, \lambda) \sim \left(\frac{3}{8\pi}\right)^{1/2} \frac{(n-2)^{5/2}}{\lambda} e^{-(8/3)[\lambda^2/(n-2)]}$$
$$= \{2\pi[\log(n-2) + w(n)]\}^{-1/2} e^{-2w(n)}.$$

This completes the proof of the theorem.

COROLLARY 1. If $\lambda = ((\frac{3}{4} + \epsilon)(n-2) \log(n-2))^{1/2}$, where ϵ is any positive constant, then almost all tournaments T_n have the property that

$$|r_{ij} - \frac{1}{4}(n-2)| \leq \lambda$$

for every pair of distinct nodes p_i and p_j.

Proof. Let $N(k)$ denote the number of tournaments T_n such that the inequality $|r_{ij} - \frac{1}{4}(n-2)| > \lambda$ holds for exactly k ordered pairs of distinct nodes p_i and p_j. Then

$$0 \leq \frac{2^{\binom{n}{2}} - N(0)}{2^{\binom{n}{2}}} = \frac{N(1) + N(2) + \cdots + N(n(n-1))}{2^{\binom{n}{2}}}$$

$$\leq \frac{0 \cdot N(0) + 1 \cdot N(1) + \cdots + n(n-1) \cdot N(n(n-1))}{2^{\binom{n}{2}}} = E(n, \lambda).$$

If $w(n) = \frac{4}{3}\epsilon \log(n-2)$ in Theorem 18, then

$$E(n, \lambda) \sim (2\pi(1 + \frac{4}{3}\epsilon) \log(n-2))^{-1/2}(n-2)^{-(8/3)\epsilon},$$

and

$$\lim_{n\to\infty} E(n, \lambda) = 0.$$

Therefore,

$$\lim_{n\to\infty} \left(1 - N(0)\cdot 2^{-\binom{n}{2}}\right) = 0,$$

and the corollary is proved.

COROLLARY 2. Almost all tournaments have diameter three.

Notice that, if the definition of the diameter of a tournament involved ordered pairs of distinct nodes only, then we could assert that almost all tournaments have diameter two.

Exercises

1. Supply the details omitted in the proof of Theorem 18.

2. Prove that $d \leq n - 1$ for any strong tournament T_n with at least four nodes.

3. Prove that $d \leq \max\{3, s_n - s_1 + 2\}$, where s_n and s_1 denote the largest and smallest score in a strong tournament T_n.

13.   The Powers of Tournament Matrices

Let M denote a square matrix with nonnegative elements. If there exists an integer t such that M^t has only positive elements, then M is *primitive* and the least such integer t is called the *exponent* of M. We may assume that the elements of M are zeros and ones, and we shall use Boolean arithmetic in calculating the powers of M. A well-known result due to Wielandt (1950) states that, if the primitive n by n matrix M_n has exponent e, then $e < (n-1)^2 + 1$.

The *directed graph* D_n defined by a matrix $M_n = [a_{ij}]$ of zeros and ones consists of n nodes $p_1, p_2, \cdots, p_n$ such that an arc $\overrightarrow{p_ip_j}$ goes from p_i to p_j if and only if $a_{ij} = 1$. The (i, j)-entry in M_n^t is one if and only if there exists a path $P(p_i, p_j)$ of length t in the graph D_n. (We cannot assume, however, that all the elements in $P(p_i, p_j)$ are distinct.)

Several authors have used properties of directed graphs to obtain results on primitive matrices and their exponents. [See, for example, the expository article by Dulmage and Mendelsohn (1965).] Most of these results are unnecessarily weak or complicated when applied to tournament matrices. Moon and Pullman (1967) showed that some fairly sharp results for primitive tournament matrices could be deduced as simple consequences of Theorem 3. If a tournament matrix M_n has a certain property (for example, if it is primitive), we shall say that the corresponding tournament T_n has the same property.

THEOREM 19. If the tournament T_n is irreducible and $n \geq 5$, then T_n is primitive; if d and e denote the diameter and exponent of T_n, then $d \leq e \leq d + 3$.

Proof. It is obvious that $d \leq e$ if T_n is primitive; therefore, we need only show that there exists a path $P(p, q)$ of length $d + 3$ for any ordered pair of nodes p and q of T_n. There exists at least one nontrivial path $P(p, q)$, since T_n is irreducible and $n \neq 1$. Let $P_1(p, q)$ be the shortest such path and let l denote its length. Then $3 \leq d - l + 3 \leq n + 2$, since $0 \leq l \leq d \leq n - 1$. If $3 \leq d - l + 3 \leq n$, then there exists a cycle $P_2(p, p)$ of length $d - l + 3$ by Theorem 3. Then the path $P(p, q) = P_2(p, p) + P_1(p, q)$ has length $d + 3$. If $d - l + 3 = n + h$, where $h = 1$ or 2, then $3 \leq n + h - 3 \leq n - 1$ since $n \geq 5$. Hence, there exist cycles $P_3(p, p)$ and $P_4(p, p)$ of lengths 3 and $n + h - 3$. The path

$$P(p, q) = P_3(p, p) + P_4(p, p) + P_1(p, q)$$

has length $3 + (n + h - 3) + l = d + 3$. This completes the proof of the theorem.

COROLLARY. A tournament T_n is primitive if and only if $n \geq 4$ and T_n is irreducible.

This result, apparently first stated by Thompson (1958), follows from Theorem 19 and the obvious fact that a primitive tournament must be irreducible. (It is easily verified that the only primitive tournament with fewer than five nodes is the strong tournament T_4; it has exponent nine.)

COROLLARY. If T_n ($n \geq 5$) is a primitive tournament with exponent e, then $3 \leq e \leq n + 2$.

There are six irreducible tournaments T_5 (see the Appendix) and they realize the exponents four, six, and seven. However, there are no gaps in the exponent set of larger primitive tournaments.

THEOREM 20. If $3 \leq e \leq n + 2$, where $n \geq 6$, then there exists a primitive tournament T_n with exponent e.

Proof. Let n and k be integers such that $3 \leq k \leq n - 1$ and $n \geq 6$. Consider the tournament T_n defined as follows: The arcs $\overrightarrow{p_1 p_n}$, $\overrightarrow{p_n p_{n-1}}$, $\cdots$,

$\overrightarrow{p_3 p_2}$, $\overrightarrow{p_2 p_1}$ and all arcs $\overrightarrow{p_j p_i}$ where $k \le i < j \le n$ are in T_n; the remaining arcs are all oriented toward the node with the larger subscript. (The structure of T_n is illustrated in Figure 5.) This tournament contains a spanning cycle so it is irreducible and hence primitive.

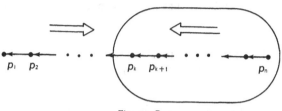

Figure 5

If $k \ge 3$, then it is easy to see that the diameter of T_n is k. The exponent of T_n is not k or $k + 1$, however, since there are no paths $P(p_k, p_1)$ of these lengths; neither is it $k + 2$, since there is no path $P(p_{k+1}, p_1)$ of this length. Therefore, the exponent of T_n is $k + 3$, by Theorem 19. This shows that there exists a primitive tournament T_n with exponent e if $6 \le e \le n + 2$ and $n \ge 6$.

If $k = 2$ or 3 let T_n' differ from T_n in that the arc joining p_{k+1} and p_n is oriented toward p_n. It is not difficult to verify that T_n' has exponent $k + 2$. Therefore, to complete the proof of the theorem, we need only show that there exist primitive tournaments T_n with exponent three when $n \ge 6$. If $n = 6$, let T_n be the tournament whose matrix is

$$\begin{vmatrix} 0 & 1 & 0 & 0 & 0 & 1 \\ 0 & 0 & 1 & 0 & 1 & 0 \\ 1 & 0 & 0 & 1 & 0 & 0 \\ 1 & 1 & 0 & 0 & 1 & 0 \\ 1 & 0 & 1 & 0 & 0 & 1 \\ 0 & 1 & 1 & 1 & 0 & 0 \end{vmatrix};$$

if $n > 6$, let T_n be the regular tournament R_n that was defined in Section 5. A simple and direct argument shows that these tournaments have exponent three. This completes the proof of the theorem.

The *index (of maximum density)* of a square matrix M is the least integer k such that the number of nonzero entries in M^k is maximized. If M is primitive, then its exponent and index are equal.

Before stating the next result, we observe that every tournament T_n has a unique decomposition into subtournaments $T^{(i)}$ $(i = 1, 2, \cdots, l)$ such that

(a) every node in $T^{(j)}$ dominates every node in $T^{(i)}$ if $1 \le i < j \le l$;
(b) every subtournament $T^{(i)}$ is either irreducible or transitive;
(c) no two consecutive subtournaments $T^{(i)}$ and $T^{(i+1)}$ are both transitive.

When T_n is itself irreducible or transitive, then $l = 1$ and $T_n = T^{(1)}$.

THEOREM 21. Let $k = $ index (M) and $k_i = $ index (M_i), where M and M_i denote the matrices of the tournament T_n and $T^{(i)}$. Then

$$k \leq \max (k_i : i = 1, 2, \cdots, l).$$

Proof. We may suppose that the matrix M has the form

$$M = \begin{vmatrix} M_1 & & & & & 0 \\ & M_2 & & & & \\ & & \cdot & & & \\ & & & \cdot & & \\ & & & & \cdot & \\ 1 & & & & & M_l \end{vmatrix},$$

where the entries above and below the diagonal blocks are zeros and ones, respectively. Then

$$M^t = \begin{vmatrix} M_1{}^t & & & & & 0 \\ & M_2{}^t & & & & \\ & & \cdot & & & \\ & & & \cdot & & \\ & & & & \cdot & \\ 1 & & & & & M_l{}^t \end{vmatrix}, \tag{1}$$

for $t = 1, 2, \cdots$.

Let $|B|$ denote the number of nonzero entries in the matrix B, and let $J = \{j : T^{(j)} \text{ is transitive}\}$. It follows from (1) that

$$|M^T| = \sum_{j \in J} |M_j{}^t| + \sum_{i \notin J} |M_i{}^t| + K,$$

where the constant K denotes the number of ones below the diagonal blocks of M.

If $m = \max (k_i : i = 1, 2, \cdots, l)$, then $|M_i{}^t| \leq |M_i{}^m|$ for all $i \notin J$ if $t \geq 1$, since every subtournament $T^{(i)}$ $(i \notin J)$ is either primitive or a 3-cycle. Furthermore, $|M_j{}^t| \leq |M_j{}^m|$ for all $j \in J$ if $t \geq m$, since $|M_j{}^{t+1}| \leq |M_j{}^t|$ for all $j \in J$ if $t \geq 1$. Therefore, $|M^t| \leq |M^m|$ if $t \geq m$, and the theorem is proved. Strict equality holds in Theorem 21 if no transitive subtournament $T^{(j)}$ has two or more nodes.

COROLLARY. If η denotes the maximum number of nodes in any of the irreducible subtournaments $T^{(i)}$, then

$$k \leq \begin{cases} 1 & \eta \leq 3, \\ 9 & \eta = 4, \\ \max (\eta + 2, 9) & \text{if} \quad \eta > 4, \\ n + 2 & \eta \geq 7. \end{cases}$$

This follows from the second corollary to Theorem 19 and the fact that the irreducible tournaments T_1, T_3, and T_4 have index one, one, and nine.

Since we are using Boolean arithmetic, there are only finitely many distinct matrices among the powers of a given matrix M of zeros and ones. The *index of convergence* and the *period* of M are the smallest positive integers $\gamma = \gamma(M)$ and $\rho = \rho(M)$ such that $M^{\rho+\gamma} = M^\gamma$. If M is primitive, then γ equals the exponent and the index of maximum density of M, and $\rho = 1$.

Let α denote the maximum number of nodes in any of the transitive subtournaments $T^{(j)}$ in the decomposition of T_n and let $\beta = \max (k_i : i \notin J)$. (We adopt the convention that the maximum of an empty set is zero.)

THEOREM 22.   If M is the matrix of the tournament T_n, then

$$\gamma(M) = \max (\alpha, \beta) \leq n + 2$$

and

$$\rho(M) = \begin{cases} 3 & \text{if some irreducible subtournament } T^{(i)} \text{ has three nodes,} \\ 1 & \text{otherwise.} \end{cases}$$

Proof.   It follows from (1) that

$$\gamma(M) = \max (\gamma(M_i) : i = 1, 2, \cdots, l).$$

However,

$$\alpha = \max (\gamma(M_j) : j \in J)$$

and

$$\beta = \max (k_i : i \notin J) = \max (\gamma(M_i) : i \notin J).$$

This suffices to establish the first part of the theorem; the second part is obvious.

THEOREM 23.   If $n \geq 16$ and $1 \leq k \leq n + 2$, then there exists a tournament T_n with index k.

Proof.   The transitive tournament T_n has index one for all n. In view of Theorem 20, it remains only to exhibit a tournament T_n with index two for $n \geq 16$.

If $n \geq 16$, let $n = 3h + r$, where $r = 16$, 17, or 18, and h is an integer. Let T_n be the tournament that can be decomposed in the manner described earlier into the subtournaments $T^{(i)}$ $(i = 1, 2, \cdots, h + 2)$, where $T^{(1)}$ is the transitive tournament T_{r-7}, $T^{(2)}$ is the tournament T_7 in which $p_i \rightarrow p_j$ if and only if $j - i$ is a quadratic residue modulo seven, and the tournament $T^{(i)}$ is a 3-cycle, for $i = 3, 4, \cdots, h + 2$. It is a simple exercise to verify that T_n has index two. (Notice that the 3-cycles $T^{(i)}$ $(i \geq 3)$ have no affect on the index of T_n; they merely extend the basic construction to tournaments with an arbitrary number n of nodes, where $n \geq 16$.)

Exercises

1. If s_n and s_1 denote the largest and smallest scores in a primitive tournament T_n ($n \geq 5$), then show that the exponent satisfies the inequality $e \leq 5 + s_n - s_1$.

2. Determine the exponents of the primitive tournaments with at most five nodes.

3. Supply the details omitted in the proofs of Theorems 20 and 23.

4. Verify Equation (1) in the proof of Theorem 21.

5. Determine whether there exist tournament matrices M_n with index two when $n < 16$.

6. Prove that almost all tournament matrices M_n are primitive with exponent three.

14.   Scheduling a Tournament

If n players are to participate in a round-robin tournament T_n, then the problem arises of scheduling the matches between the players. When n is even, the $\frac{1}{2}n(n-1)$ matches can be split into $n-1$ rounds of $\frac{1}{2}n$ matches each in such a way that every pair of players meet exactly once; when n is odd, an extra round is needed. We may suppose that n is even, for if n is odd we may introduce an imaginary player P and let each player sit out the round in which he is matched with P. We now outline a simple scheme Reisz (1859) gave for scheduling the matches of a round-robin tournament between an even number n of players.

First write the pairs $(1, 2)$, $(1, 3)$, $\cdots$, $(1, n)$ in the first column of an $n-1$ by $n-1$ table. Then, for $i = 2, 3, \cdots, n-1$, write the pairs $(i, i+1)$, $(i, i+2)$, $\cdots$, (i, n) in the ith column, beginning in the first row below the row containing $(i-1, i)$ that does not already contain an i or an $i+1$, and returning to the top of the column when the bottom is reached. Whenever this rule would place an entry (i, j) in a position in a row that already contains an i or a j, leave the position vacant and place (i, j) in the next admissible position. When all the pairs (i, j) with $1 \leq i < j \leq n$ have been entered, then the pairs in the rth row indicate the matches of the rth round.

The following table illustrates Reisz's construction for the case $n = 8$.

(1, 2)		(3, 7)	(4, 6)	(5, 8)		
(1, 3)	(2, 8)		(4, 7)	(5, 6)		
(1, 4)	(2, 3)			(5, 7)	(6, 8)	
(1, 5)	(2, 4)	(3, 8)			(6, 7)	
(1, 6)	(2, 5)	(3, 4)				(7, 8)
(1, 7)	(2, 6)	(3, 5)	(4, 8)			
(1, 8)	(2, 7)	(3, 6)	(4, 5)			

The table can be put in the following, more compact, form.

(1, 2)	(3, 7)	(4, 6)	(5, 8)
(1, 3)	(2, 8)	(4, 7)	(5, 6)
(1, 4)	(2, 3)	(5, 7)	(6, 8)
(1, 5)	(2, 4)	(3, 8)	(6, 7)
(1, 6)	(2, 5)	(3, 4)	(7, 8)
(1, 7)	(2, 6)	(3, 5)	(4, 8)
(1, 8)	(2, 7)	(3, 6)	(4, 5)

This construction matches players i and j, where $1 \leq i, j \leq n - 1$, in round r if

$$i + j \equiv r + 2 \pmod{n - 1},$$

and it matches players k and n in round r if

$$2k \equiv r + 2 \pmod{n - 1},$$

for $r = 1, 2, \cdots, n - 1$. If the $r + 2$ in these congruences is replaced by r, the only effect on the schedule is that the rounds are numbered differently. König (1936, p. 157) gave this construction [see also Freund (1959) and Lockwood (1962)]. Slightly different schemes for generating similar schedules have also been given by Lockwood (1936), Kraitchik (1950, p. 230), and Ore (1963, p. 50). This problem is only a special case of the more general problem of finding factors of a graph that satisfy certain conditions.

Some more complicated problems on the design of other types of tournaments where more than two players meet at a time have been treated by Kraitchik, Watson (1954), Scheid (1960), Gilbert (1961), and Yalavigi (1963). There are also a number of design problems connected with various modifications of the method of paired comparisons where not every pair of objects is compared or where there is more than one judge. Certain conditions on balance and symmetry are usually imposed. Material on these problems and other references may be found in Kendall (1955), Bose (1956), Tietze (1957), and David (1963).

Exercises

1. The following array illustrates for the case $n = 7$ a construction Kraitchik gives for scheduling a tournament between an odd number of players. (Players i and j meet in round r if the pair ij occurs in row r and player k sits out round r if k occurs in the rth row of column 1.)

$$
\begin{array}{cccc}
1 & 72 & 63 & 54 \\
5 & 46 & 37 & 21 \\
2 & 13 & 74 & 65 \\
6 & 57 & 41 & 32 \\
3 & 24 & 15 & 76 \\
7 & 61 & 52 & 43 \\
4 & 35 & 26 & 17 \\
\end{array}
$$

Extend this construction to the general case. Notice that a schedule for a tournament on 8 players can be obtained from this by writing an 8 next to the numbers in the first column. Is the resulting schedule different from the schedule obtained by Reisz's construction?

2. Describe how the diagram in Figure 6 can be used to schedule a tournament between seven players. How can the construction be generalized?

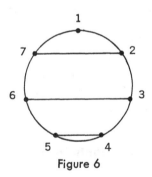

Figure 6

3. A tennis match is played between two teams. Each player plays one or more members of the other team. Any two members of the same team have exactly one opponent in common and no two members of the same team play all the members of the other team between them. Prove that, if two players on different teams do not play each other, then they have the same number of opponents. Deduce from this that all the players have the same number of opponents and that, if this number is n, then there are $n^2 - n + 1$ players on each team. [Boyd (1961).]

4. A tennis match is played between two teams A and B. Each member of A plays at least one member of B, and no member of B plays every member of A. Prove that there exist players a_1, a_2, b_1, and b_2 such that a_i and b_j have played each other if and only if $i = j$ for $i, j = 1, 2$. [See McKay (1966).]

5. The members of a bridge club participated over a period of several days in a tournament that satisfied the following conditions:

(a) Each pair of members appeared together at exactly one day's play.
(b) For any two days' play there was exactly one member who played on both days.
(c) At least four players were scheduled to play on each day.
(d) The president, vice president, secretary, and treasurer were the only members scheduled to play on the first day.
    How many members did the club have, how many days did the play last, and how could the tournament have been scheduled? [Mendelsohn (1953).]

6. The nine members of a whist club want to arrange a tournament among themselves. There are nine chairs in their club room, four at each of two tables and an extra one for the person who sits out each round. Is it possible to arrange the tournament so that all three of the following conditions are satisfied?

(a) Every two members play together as partners once.
(b) Every two members play together as opponents twice.
(c) Each member spends one round in each chair.

Consider the analogous problem for clubs with five and thirteen members. [See Watson (1954) and Scheid (1960).]

7. When is it possible to schedule a round robin tournament between n chess players in such a way that all players alternate in playing the white and black pieces? [Tietze (1957).]

15.   Ranking the Participants in a Tournament

The simplest way of ranking the participants in a tournament is according to the number of games they have won. This, however, will lead to ties, except when the tournament is transitive. One could consider the subtournament determined by all players who have the same score and then try to rank these players on the basis of their performance within this subtournament, but this will not necessarily remove all the ties. In fact, there is no reason to expect that all ties can be removed. For example, if the result of a tournament between three players is a 3-cycle, the most plausible conclusion is that they are of equivalent strength.

Various schemes for ranking and comparing the participants in a tournament have been proposed, none of which is entirely satisfactory. Zermelo

(1929) developed a method based on the maximum likelihood principle. [The same method was rediscovered later by Bradley and Terry (1952) and Ford (1957).] We shall describe his procedure only for the case of a round-robin tournament, although there is no difficulty in extending it to more general situations where the number of encounters between various players is arbitrary.

We assume that each player p_i has a positive strength w_i such that in the encounter between p_i and p_j the probability that p_i will win is given by $w_i/(w_i + w_j)$. (We assume the strengths are normalized so that their sum is one.)

If we assume that the outcomes of the various matches are independent, then the probability that a particular tournament T_n is obtained is given by the formula

$$Pr(T_n : w_1, w_2, \cdots, w_n) = \prod_{i<j} \left(\frac{w_i}{w_i + w_j}\right)^{a_{ij}} \left(\frac{w_j}{w_i + w_j}\right)^{a_{ji}}$$

$$= \frac{\prod_i w_i^{s_i}}{\prod_{i<j} (w_i + w_j)},$$

where $a_{ij} = 1$ or 0 according as p_i does or does not beat p_j and s_i is the number of matches won by p_i. The problem now is to determine the value of the w_i's that maximize the probability of obtaining the tournament T_n.

If the players can be split into two nonempty classes A and B such that every player in A beats every player in B, then it is obvious that every player in A should be ranked ahead of every player in B, but there exist no nonzero maximizing values of the w_i's for the players in B (see Exercise 1). Therefore, we assume that T_n is irreducible.

If T_n is irreducible, then it can be shown that there exists a unique set of positive strengths ($w_i : i = 1, 2, \cdots, n$) that maximize $Pr(T_n : w_1, w_2, \cdots, w_n)$. Upon taking partial derivatives of the logarithm of the likelihood function, we see that the maximizing strengths satisfy the equations

$$\frac{s_i}{w_i} = \sum_j' \frac{1}{w_i + w_j} \tag{1}$$

and

$$\sum_{i=1}^n w_i = 1, \tag{2}$$

where for each i the first sum is over all j not equal to i.

There is no direct way of solving these equations in general, but an iterative scheme can be used. For any trial solution, say $w_i^{(0)} = 1/n$, let

$$w_i^{(1)} = \frac{s_i}{\sum_j' (w_i^{(0)} + w_j^{(0)})^{-1}},$$

for $i = 1, 2, \cdots, n$. If this procedure is repeated, the trial solutions will converge (slowly) to a solution of (1) if T_n is irreducible [see David (1963)]. This solution may not satisfy (2), but since the equations are homogeneous this can be remedied by multiplying through by the appropriate constant.

We now illustrate Zermelo's method on the irreducible tournament T_6 in Figure 7. We would expect players 3, 4, and 5 to have equal strengths, but it is not obvious what the relative strengths of the other players should be.

We find that $w_1 = w_2 = .38785$, $w_3 = w_4 = w_5 = .06580$, and $w_6 = .02689$ is an approximate solution to Equations (1) and (2). Notice that this gives exactly the same ranking as would be obtained simply by considering the score of each player. This is always the case for ordinary round-robin tournaments, as pointed out by Zermelo and Ford (see Exercise 2). Hence, all we really achieve in this case are the maximum likelihood estimates for the relative strengths of the players.

The preceding method of ranking the players in a tournament and assigning their relative strengths takes into account only the number of matches won by each player. A method due to Wei (1952) and Kendall (1955) also takes into account the quality of the opponents in the matches won by each player. For example, it seems plausible that Player 1 is stronger than Player 2 in the tournament in Figure 7, even if they do have the same number of

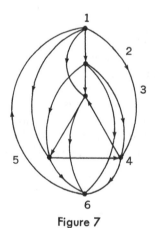

Figure 7

wins, because Player 1 beats Player 2. Also, the relative position of the sixth player is not clear, because although he wins only one match he wins against one of the strongest players. We now illustrate the Kendall-Wei method, or rather a slight simplification of it, for this tournament.

First assign to each player the number of matches he has won. This gives the initial strength vector

$$w_1 = (4, 4, 2, 2, 2, 1).$$

Next assign to each player the sum of the initial strengths of the players he has beaten. This gives the second vector

$$w_2 = (10, 7, 3, 3, 3, 4).$$

The next four strength vectors, defined in a similar way, are

$$w_3 = (16, 13, 7, 7, 7, 10),$$
$$w_4 = (34, 31, 17, 17, 17, 16),$$
$$w_5 = (82, 67, 33, 33, 33, 34),$$
$$w_6 = (166, 133, 67, 67, 67, 82).$$

Notice that the first player has the largest strength now and that the sixth player has the third largest strength. It seems unlikely that further repetitions of this process will alter the relative positions of the players.

In the general case, the ith strength vector is given by

$$w_i = M^i e, \tag{3}$$

where M is the matrix of the tournament being considered and e is a column vector of 1's. It follows from a theorem of Frobenius (1912) that, if the matrix M is primitive, then

$$\lim_{i \to \infty} \left(\frac{M}{\lambda}\right)^i e = y, \tag{4}$$

where λ is the unique positive characteristic root of M with the largest absolute value and y is a positive characteristic vector of M corresponding to λ. Therefore, in view of the corollary to Theorem 19, the normalized vector y can be taken as the vector of relative strengths of the players in T_n if T_n is irreducible and $n \geq 4$.

The characteristic polynomial of the matrix of the tournament in Figure 7 is $\lambda^6 - 5\lambda^3 - 6\lambda^2 - 6\lambda - 2$. The dominant root of this is approximately $\lambda = 2.1106295$. Upon calculating the corresponding characteristic vector and normalizing, we find that the relative strengths assigned to the players by this method are approximately $w_1 = .27899$, $w_2 = .23179$, $w_3 = w_4 = w_5 = .11901$, and $w_6 = .13218$. It would seem that these strengths are somewhat more realistic than those obtained by the maximum likelihood method.

Kendall's original version also applies to more general situations. He first assigns to each player the number of matches he has won, plus ½ for tieing with himself. At each subsequent stage of adjusting the strengths, he assigns to each player ½ his old strength plus the sum of the strengths of the players he beats. Thompson (1958) showed that the ½ is arbitrary and that, if the ½ is replaced by r, then the final ranking is independent of r if $r > 0$. If this method is used, then the matrix M in Equations (3) and (4) is replaced by $M + rI$. This is still somewhat arbitrary and it seems that the only reason for the rI is to ensure that the matrix used in (3) and (4) is primitive. But the matrix of an irreducible tournament T_n is itself primitive if $n \geq 4$ so that there is no need for the rI in this case, especially since it has no effect upon the final relative strength vector obtained.

Katz (1953) and Thompson (1958) proposed another method for comparing the participants in a tournament. If M is the matrix of the tournament, they let the vector of relative strengths be proportional to

$$(M + rM^2 + r^2M^3 + \cdots) e = M(I - rM)^{-1}e,$$

where r is some positive constant for which the series converges (that is to say, $r < \lambda^{-1}$ where λ is the dominant characteristic root of M). Thompson shows that the normalized relative strengths given by the vector $M(I - \lambda^{-1}M)^{-1}e$ are the same as those given by the Kendall-Wei method.

Additional material on the problem of ranking a collection of objects on the basis of binary comparisons between them may be found, for example, in Brunk (1960), Slater (1961), Hasse (1961), Buhlman and Huber (1963), Huber (1963b), Gridgeman (1963), David (1963), Thompson and Remage (1964), and Kadane (1966). Good (1955) has treated a related problem concerning the grading of chess players.

Exercises

1. Show that there is no positive solution of the equation $\Sigma_{i=1}^{n} w_i = 1$ that maximizes $Pr(T_n : w_1, w_2, \cdots, w_n)$ if the tournament T_n is reducible.

2. Suppose $(w_i : i = 1, 2, \cdots, n)$ satisfies Equations (1) and (2). Show that if $s_i < s_j$, then

$$0 < \sum_k{}' \frac{w_j}{w_j + w_k} - \sum_k{}' \frac{w_i}{w_i + w_k} = (w_j - w_i) \sum_k \frac{w_k}{(w_i + w_k)(w_j + w_k)},$$

so $w_i < w_j$ also.

3. Thompson raises the following questions about his method of ranking the participants in a tournament:

(a) Does the final ranking depend on the choice of r?
(b) Can the method lead to a tie between an even number of players?

4. Slater (1961) suggests that the participants in a tournament T_n be ranked in such a way as to minimize the number of *upsets*, that is, the number of matches in which the losing player is ranked ahead of the winning player. If the players are ranked in this way, does this imply that the players are ranked according to the number of matches they have won?

5. The relative weakness vector of the participants in a tournament T_n may be defined as the vector obtained by applying the Kendall-Wei method to the transpose of the matrix of T_n. Ramanujacharyula (1964) suggests that the participants be ranked according to the quotients of their relative strengths and weaknesses. Apply this method to the tournament T_6 considered in this section.

16.   The Minimum Number of Comparisons Necessary to Determine a Transitive Tournament

If one knows in advance that a dominance relation defined on a set of $n(n \geq 4)$ objects is transitive, then it is not necessary to compare all the $\binom{n}{2}$ pairs of the objects in order to determine the transitive tournament T_n that this relation defines. (For example, one might use a balance scale to rank according to weight n objects, no two of which had the same weight.) The number of comparisons necessary to rank the n objects will depend on the order in which they are compared, in general. We now give bounds for $M(n)$, the least integer M such that at most M comparisons are necessary to rank n objects according to some transitive relation.

THEOREM 24.   If $n = 2^{t-1} + r$, where $0 \leq r < 2^{t-1}$ and $n > 2$, then

$$1 + [\log_2 (n!)] \leq M(n) \leq 1 + nt - 2^t \leq 1 + n[\log_2 n].$$

Proof. The lower bound follows immediately from the observation [see Ford and Johnson (1959)] that, if M comparisons will always suffice to rank n objects, then it must be that $2^M \geq n!$; since after M comparisons we can distinguish at most 2^M alternatives and we must be able to distinguish between the $n!$ possible rankings.

The upper bound follows from a construction due to Steinhaus (1950). Compare any two of the objects at the outset. If k objects have been ranked relative to each other, compare any $(k + 1)$st object A with one of the objects, say B, ranked in the middle of these k objects. Next, compare A with one of the objects ranked in the middle of those objects already ranked above or below B, according as A dominates B or B dominates A. If this procedure is repeated, the position of A relative to the k objects already ranked will be determined after $1 + [\log_2 k]$ comparisons.

Thus, if $n = 2^{t-1} + r$, where $0 \leq r < 2^{t-1}$, the total number of matches necessary to rank the n objects is at most

$(n - 1) + [\log_2 2] + [\log_2 3] + \cdots + [\log_2 (n - 1)]$

$\quad = (n - 1) + 2(1) + 2^2(2) + 2^3(3) + \cdots + 2^{t-2} (t - 2) + r(t - 1)$

$\quad = (n - 1) + 2[1 + (t - 3) 2^{t-2}] + (n - 2^{t-1})(t - 1)$

$\quad = 1 + nt - 2^t \leq 1 + n[\log_2 n].$

Ford and Johnson (1959) have given a somewhat sharper upper bound by means of a more efficient and complicated procedure. The top two rows of Table 4 give the number of comparisons sufficient to rank n objects by the procedures of Steinhaus, and Ford and Johnson, for $n \leq 13$; the bottom row gives the lower bound for the number of comparisons necessary.

Table 4. $M(n)$, the minimum number of comparisons necessary to determine a transitive tournament T_n.

$n = 1$	2	3	4	5	6	7	8	9	10	11	12	13
0	1	3	5	8	11	14	17	21	25	29	33	37
0	1	3	5	7	10	13	16	19	22	26	30	34
0	1	3	5	7	10	13	16	19	22	26	29	33

Ford and Johnson conjecture that $M(n)$ equals the number of comparisons required by their procedure. Steinhaus (1963) conjectures that $M(n) = 1 + [\log_2 (n!)]$ if $n > 2$ and discusses recent progress on this conjecture. The first unsettled case is $n = 12$.

Kislicyn (1963) has given an asymptotic bound for the least mean number of comparisons necessary to rank n objects.

Let $M_k(n)$ denote the least integer M such that at most M comparisons are necessary to determine the kth highest ranking object in a set of n objects. It is not difficult to see that $M_1(n) = n - 1$. More generally, Kislicyn (1964) has shown that

$$M_k(n) \leq (n - 1) + \sum_{i=1}^{k-1} [\log_2 (n - i)] \tag{1}$$

if $k \leq [\frac{1}{2}(n + 1)]$. The special cases $k = 1$ and 2 of this result were obtained earlier by Schreier (1932) and Slupecki (1951).

Exercises

1. Prove inequality (1) for $k = 1, 2$.

2. A *knock-out tournament* between n players may be conducted as follows: If $n = 2^t + r$, where $1 \leq r \leq 2^t$, then $2r$ players are matched off in the first round. The 2^t players not yet defeated are then matched off in the second round. In the ith round, the 2^{t+2-i} players not yet defeated are matched off. The undefeated winner emerges after $t + 1$ rounds. What is the probability that two given players will be matched against each other in the course of a random knock-out tournament? [Hartigan (1966) has dealt with some problems of estimating the relative ranks of participants in a knock-out tournament on the basis of the known outcomes of the matches that are played. David (1963) gives more material on this type of tournament.]

3. A certain tournament consisted of five rounds, a semifinal, and a final. In the five rounds, there were 8, 6, 0, 1, 0 byes, respectively. If there were 100 entrants, then how many matches were played? [Chisholm (1948).]

4. Prove the identity

$$\left[\frac{n}{2}\right] + \left[\frac{n+1}{4}\right] + \left[\frac{n+3}{8}\right] + \left[\frac{n+7}{16}\right] + \cdots = n - 1.$$

Hint: Consider a knock-out tournament on n players in which as many players as possible are matched off in each round and the losers of these matches are eliminated from further play. [Mendelsohn (1949).]

17.   Universal Tournaments

A tournament T_N is said to be *n-universal* ($n \leq N$) if every tournament T_n is isomorphic to some subtournament of T_N. For every positive integer n, let $\lambda(n)$ denote the least integer N for which there exists an n-universal tournament T_N. (It is clear that $\lambda(n)$ is finite, since any tournament that contains disjoint copies of all the different tournaments T_n is n-universal.)

THEOREM 25

$$2^{(1/2)(n-1)} \leq \lambda(n) \leq \begin{cases} n \cdot 2^{(1/2)(n-1)} & \text{if } n \text{ is odd,} \\ \dfrac{3}{2\sqrt{2}}\, n \cdot 2^{(1/2)(n-1)} & \text{if } n \text{ is even.} \end{cases}$$

Proof. There are at least $2^{\binom{n}{2}}/n!$ different tournaments T_n (the labelings assigned to the nodes are immaterial to the problem). Hence, if T_N is n-universal, it must be that

$$2^{\binom{n}{2}}/n! \leq \binom{N}{n} \leq \frac{N^n}{n!},$$

since different tournaments T_n must be isomorphic to different subtournaments with n nodes of T_N. The lower bound for $\lambda(n)$ is an immediate consequence of this inequality.

To obtain an upper bound for $\lambda(n)$, we proceed as follows. Let R_n be any tournament with nodes $y_1, y_2, \cdots, y_n$ and let $Y_i = \{y_k : y_k \to y_i\}$ for each i. Construct a tournament H whose nodes $q_{i,A}$ are in one-to-one correspondence with the ordered pairs (i, A) where $A \subset Y_i$. If $A \subset Y_i$, $B \subset Y_j$, and $y_i \to y_j$, then $q_{i,A} \to q_{j,B}$ in H if $y_i \in B$, and $q_{j,B} \to q_{i,A}$ if $y_i \notin B$. The arcs joining nodes of the type q_{i,A_1} and q_{i,A_2} may be oriented arbitrarily. (The tournament obtained when R_n is a 3-cycle is illustrated in Figure 8. The notation used should be obvious.)

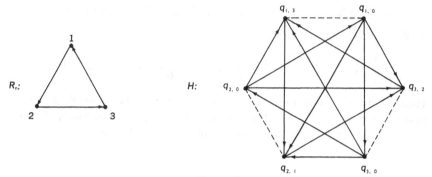

Figure 8

We now show that H is n-universal. If the nodes of a tournament T_n are $p_1, p_2, \cdots, p_n$, we set $f(p_i) = q_{i,A(i)}$, where $A(i) = \{y_k : y_k \to y_i$ in R_n and $p_k \to p_i$ in $T_n\}$. The subtournament of H determined by the nodes $f(p_i)$ is isomorphic to T_n, since, if $y_i \to y_j$ in R_n, then

$$(p_i \to p_j) \Rightarrow (p_i \in A(j)) \Rightarrow (q_{i,A(i)} \to q_{j,A(j)}) \Rightarrow (f(p_i) \to f(p_j))$$

and

$$(p_j \to p_i) \Rightarrow (p_i \notin A(j)) \Rightarrow (q_{j,A(j)} \to q_{i,A(i)}) \Rightarrow (f(p_j) \to f(p_i)).$$

Therefore,

$$\lambda(n) \leq (\text{number of nodes of } H) = 2^{|Y_1|} + 2^{|Y_2|} + \cdots + 2^{|Y_n|}.$$

To minimize this sum, let R_n be the regular tournament defined in Section 5 for which

$$|Y_1| = \cdots = |Y_n| = \tfrac{1}{2}(n-1), \qquad \text{if } n \text{ is odd,}$$

and

$$|Y_1| = \cdots = |Y_{(1/2)n}| = \tfrac{1}{2}n, \ |Y_{(1/2)n+1}| = \cdots$$
$$= |Y_n| = \tfrac{1}{2}(n-2), \qquad \text{if } n \text{ is even.}$$

Hence,

$$\lambda(n) \leq n \cdot 2^{1/2(n-1)}, \qquad \text{if } n \text{ is odd,}$$

and

$$\lambda(n) \leq \tfrac{1}{2}n \cdot 2^{(1/2)n} + \tfrac{1}{2}n \cdot 2^{(1/2)(n-2)} = \frac{3}{2\sqrt{2}} n \cdot 2^{(1/2)(n-1)}, \qquad \text{if } n \text{ is even.}$$

This completes the proof of the theorem.

Rado (1964) and de Bruijn were the first to study universal graphs; they restricted their attention to infinite graphs. Theorem 25 is closely related to a result Moon (1965) obtained for ordinary finite graphs.

Exercises

1. Verify that the graph H in Figure 8 is 3-universal.

2. Determine the exact values of $\lambda(n)$ for $n \leq 4$.

3. Obtain a result for directed graphs that is analogous to Theorem 25.

18.   Expressing Oriented Graphs as the Union of Bilevel Graphs

The rather complicated results in this section will be used to prove other results in the next two sections. In order to prove these results for tournaments, it is necessary to prove them for oriented graphs in general. (Recall that an oriented graph differs from a tournament in that not all pairs of nodes need be joined by an arc.)

Recall that an a by b bipartite tournament consists of two disjoint sets A and B containing a and b nodes, respectively, such that each node in A is joined by an arc to each node in B. We shall consider special a by b bipartite graphs $H(a, b)$ in which all the arcs are similarly oriented, say, from the nodes in A to the nodes in B. A *bilevel graph H* is any oriented graph that can be expressed as the union of disjoint special graphs $H(a_i, b_i)$; the graphs $H(a_i, b_i)$ are called the *components* of H. (We admit the possibility that one of the node sets of one of the components of H is empty.) The structure of a typical bilevel graph with four components is illustrated in Figure 9.

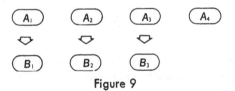

Figure 9

We are going to prove that any oriented graph with n nodes can be expressed as the union of $cn/\log n$ or fewer arc-disjoint bilevel graphs, where c is a certain constant. First, however, we prove several lemmas.

The (*total*) *degree* of a node in an oriented graph is the total number of arcs incident with it. The following lemma and some subsequent statements are valid only for large values of n; when this is clear from the context, we shall not mention this qualification explicitly.

LEMMA 1.   Let G be an oriented graph with n nodes and e arcs such that

$$\frac{n^2}{2^{2r+4}} < e \le \frac{n^2}{2^{2r+1}}$$

for some nonnegative integer r where $\frac{\log n}{3(r+3)} \ge 1$. Then G contains a special bipartite tournament $H(a, b)$ with $a = [\sqrt{n}]$ and $b = \left[\frac{\log n}{3(r+3)}\right]$ where the degrees of the nodes of A do not exceed $\frac{16n}{2^r}$ in G.

Proof. There can be at most $n/2^{r+4}$ nodes whose degrees exceed $16n/2^r$ in G, for otherwise there would be too many arcs in G. Hence, there are fewer than $n^2/2^{2r+9}$ arcs joining two such nodes. If we disregard such arcs, there will remain more than $n^2/2^{2r+5}$ arcs with the property that at least one of the nodes they join has degree at most $16n/2^r$. Let $y_1, y_2, \cdots, y_t$ denote the nodes whose degrees in G were at most $16n/2^r$ originally. Then it must be that

$$n\left(1 - \frac{1}{2^{r+4}}\right) \le t \le n.$$

If s_i denotes the number of arcs oriented away from y_i, then we may assume that

$$\sum_{i=1}^{t} s_i \ge \frac{n^2}{2^{2r+6}}.$$

(If this inequality is not satisfied, then in the special graph $H(a, b)$ all the arcs will be oriented from the nodes of B to the nodes of A, but this will not affect the result.)

The number N of sets B of b nodes for which there exists some node y_i ($1 \le i \le t$) such that y_i dominates every node of B is given by the formula

$$N = \sum_{i=1}^{t} \binom{s_i}{b}.$$

(If there are h nodes y_i that dominate every node of B, then B is counted h times in this sum.) This sum will be minimized when $t = n$ and $s_i = [n/(2^{2r+5})]$. If we can show that

$$N \ge n \binom{\left[\frac{n}{2^{2r+6}}\right]}{b} > (a - 1) \binom{n}{b}, \tag{1}$$

then it will follow from the box principle that at least one set B is counted at least a times; this means that there exists a set A of a nodes (chosen from the nodes whose degrees in G were at most $16n/2^r$ originally) and a set B of b nodes such that every node in A dominates every node in B.

It is not difficult to show (see Exercise 1) that inequality (1) does in fact hold when

$$a = [\sqrt{n}] \quad \text{and} \quad b = \left[\frac{\log n}{3(r + 3)}\right].$$

This completes the proof of Lemma 1.

LEMMA 2. Let G be an oriented graph with n nodes and e arcs such that

$$\frac{n^2}{2^{2r+3}} < e \le \frac{n^2}{2^{2r+1}}$$

for some nonnegative integer r such that $r \le 4 \log \log n$. Then G contains a bilevel graph with at least $(n \log n)/(r + 3)2^{r+11}$ arcs.

Proof. We first disregard all arcs that join two nodes whose degrees exceed $16n/2^r$; as before, there are fewer than $n^2/2^{2r+9}$ such arcs so there certainly remain more than $n^2/2^{2r+13/4}$ arcs. Let H_1 be the special subgraph of G described in Lemma 1. Since the degrees of the nodes in A_1 and B_1 are at most $16n/2^r$ and $n - 1$, respectively, it follows that the number of arcs incident with nodes of $A_1 \cup B_1$ is less than

$$n\left(\frac{16\sqrt{n}}{2^r} + \log n\right) < \frac{17n^{3/2}}{2^r}.$$

If we disregard these arcs, there still remain more than

$$\frac{n^2}{2^{2r+13/4}} - \frac{17n^{3/2}}{2^r} > \frac{n^2}{2^{2r+4}}$$

arcs. We now apply Lemma 1 again to obtain a special subgraph H_2. (The arcs in H_i need not all be oriented from nodes in A_i to nodes in B_i; it may be that they are all oriented from nodes in B_i to nodes in A_i.) We now disregard the arcs incident with the nodes of H_2 and apply Lemma 1 again. When we have repeated this procedure $[\sqrt{n}/(17 \cdot 2^{r+5})]$ times, we are left with a graph that still has more than

$$\frac{n^2}{2^{2r+13/4}} - \frac{17n^{3/2}}{2^r}\left[\frac{\sqrt{n}}{17 \cdot 2^{r+5}}\right] > \frac{n^2}{2^{2r+4}}$$

arcs. If we apply Lemma 1 once more, then the bilevel graph with components H_i, $i = 1, 2, \cdots, [\sqrt{n}/(17 \cdot 2^{r+5})] + 1$, has at least

$$\left(\left[\frac{\sqrt{n}}{17 \cdot 2^{r+5}}\right] + 1\right) \cdot [\sqrt{n}] \cdot \left[\frac{\log n}{3(r + 3)}\right] > \frac{n \log n}{(r + 3) \cdot 2^{r+11}}$$

arcs. This proves the lemma.

LEMMA 3. If G is an oriented graph with e arcs, then G contains a bilevel subgraph with at least $\frac{1}{2}\sqrt{e}$ arcs.

Proof. This is trivially true when $e = 0$; the proof of the general case is by induction. Let p be one of the nodes of largest degree d in G. We may suppose that at least $\frac{1}{2}d$ arcs are oriented away from p; use these arcs to form one component of a bilevel subgraph of G. If α denotes the number of arcs that are incident with any nodes that are joined to p, then $0 < \alpha \leq \min (d^2, e)$. We may now apply the induction hypothesis to the other $e - \alpha$ arcs and assert that G has a bilevel subgraph with at least

$$\frac{1}{2}(d + \sqrt{e - \alpha}) \geq \frac{1}{2}\sqrt{e} \qquad (2)$$

arcs.

LEMMA 4. If G is an oriented graph with e arcs, then G can be expressed as the union of fewer that $4\sqrt{e}$ arc-disjoint bilevel graphs.

Proof. Suppose that G is the union of t arc-disjoint bilevel graphs with $e_1, e_2, \cdots, e_t$ arcs, respectively, where $e_1 \leq e_2 \leq \cdots \leq e_t$. In view of Lemma 3, we may assume that

$$e_i \geq \frac{1}{2}\sqrt{e_1 + e_2 + \cdots + e_i}$$

or

$$4e_i^2 - e_i \geq e_1 + e_2 + \cdots + e_{i-1}$$

for each i. From this inequality it follows by induction that $e_i \geq \frac{1}{8}i$ for each i. Therefore

$$e = e_1 + e_2 + \cdots + e_t \geq \frac{1}{8}\binom{t+1}{2} > (\frac{1}{4}t)^2,$$

or $t < 4\sqrt{e}$.

We can now prove the following theorem due to Erdös and Moser (1964a).

THEOREM 26. There exists a constant c such that any oriented graph G with n nodes can be expressed as the union of l arc-disjoint bilevel graphs, all of which have the same n nodes, where

$$l \leq \frac{cn}{\log n}.$$

Proof. We define oriented graphs G_i and $G^{(i)}$ inductively for $i = 1$, 2, $\cdots$, $[2^{18}(n/\log n)]$. The graph $G^{(i)}$ is obtained from G by removing the arcs of $G_1, G_2 \cdots, G_i$. Let G_1 be a bilevel subgraph of G with a maximal number of arcs and let G_{i+1} be similarly defined with respect to $G^{(i)}$. (We may suppose that all the graphs G_i have n nodes.)

If there are e_i arcs in $G^{(i)}$ and if i_r is the smallest integer such that

$$e_{i_r} \le \frac{n^2}{2^{2r+1}}, \quad \text{for } r = 0, 1, \cdots, [4 \log \log n],$$

then we shall show that

$$i_{r+1} - i_r \le 2^{11} \cdot \frac{r+3}{2^{r+1}} \cdot \frac{n}{\log n}. \tag{3}$$

We may suppose that

$$e_{i_r} > \frac{n^2}{2^{2r+3}}, \quad \text{for } i_{r+1} - i_r = 0$$

otherwise. If $i_r \le j < i_{r+1}$, then

$$\frac{n^2}{2^{2r+3}} < e_j \le \frac{n^2}{2^{2r+1}}.$$

Hence, $G^{(j)}$ contains a bilevel subgraph with at least

$$\frac{n \log n}{(r+3)\, 2^{r+11}}$$

arcs, by Lemma 2. This implies that

$$e_j - e_{j+1} \ge \frac{n \log n}{(r+3)\, 2^{r+11}}.$$

If we sum this inequality over all j such that $i_r \le j < i_{r+1}$, we find that

$$\frac{n^2}{2^{2r+1}} \ge e_{i_r} - e_{i_{r+1}} = \sum (e_j - e_{j+1}) \ge (i_{r+1} - i_r) \cdot \frac{n \log n}{(r+3)\, 2^{r+11}}.$$

This implies inequality (3).

Therefore, upon removing at most

$$\sum_{0 \le r \le [4 \log \log n]} (i_{r+1} - i_r) < \frac{2^{11}\, n}{\log n} \sum_{r=0}^{\infty} \frac{r+3}{2^{r+1}} = \frac{2^{13}\, n}{\log n}$$

arc-disjoint bilevel graphs G_i, we are left with a graph G' that has at most

$$\frac{n^2}{2^{2([4 \log \log n] + 1) + 1}} < \frac{n^2}{(\log n)^5}$$

arcs. But G' is the union of fewer than $4n/(\log n)^{5/2}$ arc-disjoint bilevel graphs by Lemma 4. This completes the proof of Theorem 25.

Exercises

1. Prove that inequality (1) holds when $a = [\sqrt{n}]$ and $b = \left[\dfrac{\log n}{3(r+3)}\right]$.

2. Verify inequality (2).

19.   Oriented Graphs Induced by Voting Patterns

Suppose that m voters each rank n objects $1, 2, \cdots, n$ in order of preference. Their preferences may be represented by an m by n matrix M, each row of which is a permutation of the objects $1, 2, \cdots, n$. The collective preferences of these voters induce an oriented graph H_n with n nodes $p_1, p_2, \cdots, p_n$ in which the arc $\overrightarrow{p_i p_j}$ goes from node p_i to node p_j if and only if i precedes j in a majority of the m rows of the matrix M. This graph will be a tournament, unless m is even and some ties occur; then certain nodes will not be joined by an arc.

Even though each voter's preferences are transitive, the collective preferences determined by the majority rule need not be transitive. For example, three voters who rank three objects according to the preference matrix

$$M = \begin{vmatrix} 1 & 2 & 3 \\ 2 & 3 & 1 \\ 3 & 1 & 2 \end{vmatrix}$$

induce a 3-cycle that is certainly not transitive.

McGarvey (1953) has shown that at most $n(n - 1)$ voters are necessary to induce any oriented graph H_n. He associates a pair of voters with each arc in the graph. The two voters associated with the arc $\overrightarrow{p_i p_j}$ rank the n objects in the orders $(i, j, 1, 2, \cdots, n)$ and $(n, n - 1, \cdots, 1, i, j)$. Each pair of voters thus determines one arc $\overrightarrow{p_i p_j}$, since the preferences of the remaining voters cancel out with respect to i and j.

Stearns (1959) showed, by a more complicated construction, that at most $n + 2$ voters are necessary to induce any oriented graph H_n; he also showed that at least $\frac{1}{2} \log 3 \, (n/\log n)$ voters are necessary in some cases. This lower bound is combined with an upper bound due to Erdös and Moser (1964a) in the following result.

THEOREM 27. If $m(n)$ denotes the least integer m such that at most m voters are necessary to induce any oriented graph H_n $(n > 1)$, then there exist constants c_1 and c_2 such that

$$\frac{c_1 \, n}{\log n} < m(n) < \frac{c_2 \, n}{\log n}.$$

Proof. If no more than m voters are necessary to induce each of the $3^{\binom{n}{2}}$ oriented graphs H_n, then it must be that

$$3^{\binom{n}{2}} \leq (n! + 1)^m,$$

since each of the m possible voters can vote in one of $n!$ ways or not vote at all. The lower bound is obtained by solving for m.

We assert that any bilevel graph H can be induced by two voters. For, suppose that H has three components and that A_1, A_2, A_3 and B_1, B_2, B_3 are the node sets of the components. If

$$A_i = \{a_{ij} : j = 1, 2, \cdots, j_i\} \quad \text{and} \quad B_i = \{b_{ik} : k = 1, 2, \cdots, k_i\},$$

then H can be induced by the preference matrix

$$\begin{vmatrix} a_{11}a_{12}\cdots a_{ij_1}b_{11}b_{12}\cdots b_{1k_1}a_{21}a_{22}\cdots a_{2j_2}b_{21}b_{22}\cdots b_{2k_2}a_{31}a_{32}\cdots a_{3j_3}b_{31}b_{32}\cdots b_{3k_3} \\ a_{3j_3}\cdots a_{32}a_{31}b_{3k_3}\cdots b_{32}b_{31}a_{2j_2}\cdots a_{22}a_{21}b_{2k_2}\cdots b_{22}b_{21}a_{1j_1}\cdots a_{12}a_{11}b_{1k_1}\cdots b_{12}b_{11} \end{vmatrix}.$$

This construction can easily be extended to the general case.

Therefore, if an oriented graph H_n can be expressed as the union of l arc-disjoint bilevel graphs with n nodes, then H_n can be induced by $2l$ voters. The upper bound for $m(n)$ now follows from Theorem 26.

Exercises

1. Prove the following results about the least integer $g = g(n)$ such that at most g voters are necessary to induce any tournament T_n, where $n \geq 2$.

(a) $g(n)$ is always odd.
(b) $g(3) = g(4) = g(5) = 3$.
(c) $g(n + 1) \leq g(n) + 2$.
(d) $m(n) \leq 2g(n)$. [Stearns (unpublished).]

2. H. Robbins raised the following question: If an odd number m of voters rank n objects randomly and independently, then what is the probability $p(m, n)$ that the tournament T_n they induce is transitive? Is it true that

$$\lim_{m \to \infty} p(m, n) = \frac{n!}{2^{\binom{n}{2}}}$$

for each fixed value of n? [See DeMeyer and Plott (1967).]

3. Characterize those oriented graphs that can be induced by m voters, for $m = 1, 2, 3$.

20. Oriented Graphs Induced by Team Comparisons

Two teams of players can be compared by matching each player on one team against each player on the other team. If the players on Team A collectively win more games than they lose against players on Team B, then we say that Team A is stronger than Team B (symbolically, $A > B$).

We admit the possibility of draws, both between individual players and between teams. We denote both the names and the strengths of players by a single number and we assume that the stronger player wins in any match between two players. [Steinhaus and Trybula (1959) have mentioned a possible industrial application of this method of comparing two samples of objects.]

Let N players $p_1, p_2, \cdots, p_N$ be split into n $(n > 1)$ nonempty teams $T_1, T_2, \cdots, T_n$ and suppose that every team is compared with every other team. This induces an oriented graph H_n with n nodes $t_1, t_2, \cdots, t_n$ in which the arc $\overrightarrow{t_i t_j}$ goes from node t_i to node t_j if and only if $T_i > T_j$. For example, the teams $T_1 = \{6, 7, 2\}$, $T_2 = \{1, 5, 9\}$, and $T_3 = \{8, 3, 4\}$ induce the 3-cycle in Figure 10.

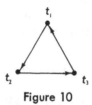

Figure 10

We now show that any oriented graph can be induced by comparing appropriate teams.

If the teams $T_1, T_2, \cdots, T_n$ generate the oriented graph H_n, let $\alpha(i, j)$ denote the number of games won minus the number of games lost by players of T_i against players of T_j. (For example, $\alpha(1, 2) = \alpha(2, 3) = \alpha(3, 1) = 1$ in the illustration in Figure 10.) We call $\alpha(i, j)$ the *net score* of T_i against T_j; notice that $\alpha(j, i) = -\alpha(i, j)$.

If w and s denote the strengths of the weakest and strongest players on the n teams, let $w_1, w_2, s_1,$ and s_2 be any numbers such that $w_1 = w_2 < w$ and $s_1 > s_2 > s$. Add two players of strength w_1 and s_1 to T_i and two players w_2 and s_2 to T_j. This will insrease $\alpha(i, j)$ by one, but the net scores between all other pairs of teams will not be affected (see Exercise 1).

We may assume that initially there are n players of equal strength, one on each team; the net scores are all zero and the induced graph has no arcs. The process described in the preceding paragraph can now be repeated as often as necessary. We conclude, therefore, that if the (nonnegative) net scores between the teams are prescribed in advance and their sum is β, then no more than $n + 4\beta$ players are necessary to realize these scores. In particular, any oriented graph H_n (for which the corresponding net scores would all be zero or one) may be induced by $n + 4 \binom{n}{2} = 2n^2 - n$ or fewer players.

The preceding argument is due to Moser; Moon and Moser (1967) gave the following sharper result that, in a sense, is the best possible.

THEOREM 28. If $N(n)$ denotes the least integer N such that at most N players are necessary to induce any oriented graph H_n $(n > 1)$, then there exist constants c_1 and c_2 such that

$$\frac{c_1 n^2}{\log n} < N(n) < \frac{c_2 n^2}{\log n}.$$

Proof. If m players can induce the graph H_n, let h_i denote the number of players that have strength i; we may suppose there exists an integer t not exceeding m such that $h_i > 0$ for $i = 1, 2, \cdots, t$ and

$$h_1 + h_2 + \cdots + h_t = m.$$

The number of solutions in positive integers to this equation is 2^{m-1} and the m players corresponding to each solution can be split into n teams in at most n^m ways. Consequently, if N or fewer players suffice to induce every oriented graph H_n, it must be that

$$3^{\binom{n}{2}} \leq \tfrac{1}{2} \sum_{m=1}^{N} (2n)^m < (2n)^N,$$

or

$$N > \frac{\log 3}{2} \frac{n(n-1)}{\log (2n)},$$

since each allocation of players determines at most one graph. This implies the lower bound of the theorem.

We assert that any bilevel graph H_n can be induced by $2n$ players, two on each team. For, suppose that H_n has three components and that A_1, A_2, A_3 and B_1, B_2, B_3 are the node sets of the components; then H_n is induced if each node in the various node sets is associated with the team indicated in the following list.

$$A_1 : (1, 10) \quad A_2 : (2, 8) \quad A_3 : (3, 6)$$
$$B_1 : (1, 9) \quad B_2 : (2, 7) \quad B_3 : (3, 5)$$

It is easy to extend this construction to the general case.

Suppose the oriented graph H_n can be expressed as the union of l arc-disjoint bilevel graphs $B^{(k)}$ $(k = 1, 2, \cdots, l)$, all of which have the same n nodes. We know that there exist teams R_{ki} of two players each such that the teams R_{ki} $(i = 1, 2, \cdots, n)$ induce the bilevel graph $B^{(k)}$ $(k = 1, 2, \cdots, l)$. We may assume that every player on any team R_{kj} is stronger than every player on any team R_{hi} $(1 \leq h < k \leq l)$. [This property can be ensured by adding a suitable constant c_k to the strength of every player on the team R_{ki} $(k = 1, 2, \cdots, l)$, if necessary.] The teams

$$T_i = \bigcup_{k=1}^{l} R_{ki}, \qquad i = 1, 2, \cdots, n,$$

have $2l$ players each and it is not difficult to see that they generate the oriented graph H_n.

Therefore, if an oriented graph H_n can be expressed as the union of l arc-disjoint bilevel graphs with n nodes, then H_n can be induced by $2ln$ players. This implies the upper bound for $N(n)$ since, by Theorem 26, we may suppose that $l \leq cn/\log n$.

There are certain rather curious aspects of this mode of comparing teams. In the example illustrated in Figure 10, the teams T_1, T_2, and T_3 were such that $T_1 > T_2$ and $T_2 > T_3$. One might expect that $T_1 \cup T_2 > T_2 \cup T_3$, and this is indeed the case. However, since $T_1 < T_3$, one might equally well expect that $T_1 \cup T_2 < T_2 \cup T_3$, and this is false.

The following example is perhaps more striking. If $A = \{2, 3, 10\}$ and $B = \{1, 8, 9\}$ then $A > B$ by 5 wins to 4. If $A_1 = A \cup \{5\}$ and $B_1 = B \cup \{4\}$, then the teams A_1 and B_1 are tied with 8 wins each. If $A_2 = A_1 \cup \{7\}$ and $B_2 = B_1 \cup \{6\}$, then $B_2 > A_2$ by 13 wins to 12. Notice that at each stage we added the stronger player to the team that was stronger originally, yet the net effect was to reverse the relative strengths of the two teams. This process can be continued. If $A_3 = A_2 \cup \{12\}$ and $B_3 = B_2 \cup \{11\}$, then A_3 and B_3 are tied with 18 wins each. Finally, if $A_4 = A_3 \cup \{14\}$ and $B_4 \cup \{13\}$, then $A_4 > B_4$ by 25 wins to 24.

Exercises

1. Verify the unproved assertion about the effect upon the net scores of adding players w_1, s_1, w_2, and s_2 to T_i and T_j.

2. Show that $N(3) = 7$.

3. The first argument in this section shows that $N(n) \leq 2n^2 - n$. Refine this argument and show by induction that $N(n) \leq n^2 + 3n - 11$ if $n \geq 3$. [Notice that this upper bound for $N(n)$ is superior to the upper bound in Theorem 28, unless n is very large.]

4. Let $w(i, j)$ and $l(i, j)$ denote the number of games won and lost by players of T_i against players of T_j. We have shown that the net scores $\alpha(i, j) = w(i, j) - l(i, j)$ for the matches between n teams can be prescribed arbitrarily. Can the win-loss ratios $w(i, j)/w(i, j)$ be prescribed arbitrarily? [See Steinhaus and Trybula (1959), Trybula (1961), and Usiskin (1964).]

5. Modify the construction used in the proof of Theorem 28 and show that any bilevel graph H_n can be induced by $2n$ players, two on each team, such that different players do not have the same strength.

6. Characterize those oriented graphs that can be induced by comparing teams with k players each, for $k = 1, 2, 3$.

7. Suppose two teams A and B are compared as follows. The ith strongest player of A is matched against the ith strongest player of B, for $i = 1, 2, \cdots,$ h, where h denotes the number of players on the smaller team. The stronger team is the one whose players win the most matches. Try to obtain a result analogous to Theorem 28 when this method of comparing teams is used.

21. Criteria for a Score Vector

The following result was first proved by Landau (1953); the proof we give here is due to Ryser (1964). [See also Alway (1962a) and Fulkerson (1966).]

THEOREM 29. A set of integers $(s_1, s_2, \cdots, s_n)$, where $s_1 \leq s_2 \leq \cdots \leq s_n$, is the score vector of some tournament T_n if and only if

$$\sum_{i=1}^{k} s_i \geq \binom{k}{2},\tag{1}$$

for $k = 1, 2, \cdots, n$ with equality holding when $k = n$.

Proof. Any k nodes of a tournament are joined by $\binom{k}{2}$ arcs, by definition. Consequently, the sum of the scores of any k nodes of a tournament must be at least $\binom{k}{2}$. This shows the necessity of (1).

The sufficiency of (1) when $n = 1$ is obvious. The proof for the general case will be by induction. Let j and k be the smallest and largest indices less than n such that $s_j = s_{s_n} = s_k$. Consider the set of integers $(s_1', s_2', \cdots, s_{n-1}')$ defined as follows:

$$s_i' = s_i \quad \text{if} \quad i = 1, 2, \cdots, j-1 \quad \text{or}$$

$$i = k - (s_n - j), \cdots, k-1, k;$$

$$s_i' = s_i - 1 \quad \text{if} \quad i = j, j+1, \cdots, k-(s_n - j) - 1 \quad \text{or}$$

$$i = k+1, k+2, \cdots, n-1.$$

From this definition, it follows that

$$s_1' \leq s_2' \leq \cdots \leq s_{n-1}',$$

that $s_1' = s_i$ for s_n values of i, and that $s_i' = s_i - 1$ for $(n-1) - s_n$ values of i. Consequently,

$$\sum_{i=1}^{n-1} s_i' = \sum_{i=1}^{n} s_i - (n-1) = \binom{n-1}{2}.$$

If there exists a tournament T_{n-1} with score vector $(s'_1, s'_2, \cdots, s'_{n-1})$, then there certainly exists a tournament T_n with score vector $(s_1, s_2, \cdots, s_n)$; namely, the tournament consisting of T_{n-1} plus the node p_n, where p_n dominates the s_n nodes p_i such that $s'_i = s_i$ and is dominated by the remaining nodes. Therefore, we need only show that the inequality

$$\sum_{i=1}^{h} s'_i < \binom{h}{2} \tag{2}$$

is impossible for every integer h such that $1 < h < n - 1$ in order to complete the proof by induction.

Consider the smallest value of h for which inequality (2) holds, if it ever holds. Since

$$\sum_{i=1}^{h-1} s'_i \geq \binom{h-1}{2},$$

it follows that $s_h \leq h$. Furthermore, $j \leq h$, since the first $j - 1$ scores were unchanged. Hence,

$$s_h = s_{h+1} = \cdots = s_f$$

if we let

$$f = \max(h, k).$$

Let t denote the number of values of i not exceeding h such that $s'_i = s_i - 1$. Then it must be that

$$s_n \leq f - t. \tag{3}$$

Therefore,

$$\binom{n}{2} = \sum_{i=1}^{h} s'_i + \sum_{i=h+1}^{f} s_i + \sum_{i=f+1}^{n-1} s_i + s_n + t$$

$$< \binom{h}{2} + (f - h) s_h + \sum_{i=f+1}^{n-1} s_i + f$$

$$\leq \binom{h}{2} + h(f - h) + \sum_{i=f+1}^{n-1} s_i + f$$

$$\leq \binom{f}{2} + f(n - f) \leq \binom{n}{2}.$$

Consequently, inequality (2) cannot hold and the theorem is proved.

Exercises

1. Show that, if the scores s_i of a tournament T_n are in nondecreasing order, then $(i - 1)/2 \leq s_i \leq (n + i - 2)/2$. [Landau (1953).]

2. Prove that $\Sigma_k\, s_i \leq k[n - (k + 1)/2]$, where the sum is over any k scores of a tournament T_n. [Landau (1953).]

3. Prove that, in any tournament, there exists a node p such that for any other node q either $p \to q$ or there exists a node r such that $p \to r$ and $r \to q$. [Landau (1953).]

4. Prove that, if a tournament has no nodes with score zero, then it has at least three nodes u such that for any other node v either $v \to u$ or there exists a node w such that $v \to w$ and $w \to u$. [Silverman (1961).]

5. Let l_i denote the number of nodes that dominate p_i in the tournament T_n so that $s_i + l_i = n - 1$ for all i. Prove in at least three different ways that

$$\sum_{i=1}^{n} s_i^2 = \sum_{i=1}^{n} l_i^2.$$

6. Is it possible for two nonisomorphic tournaments to have the same score vector?

7. Prove that among the class of all tournaments with score vector $(s_1, s_2, \cdots, s_n)$, where $s_1 \leq s_2 \leq \cdots \leq s_n$, there exists a tournament T_n whose matrix has only $\Sigma' [s_i - (i - 1)]$ ones above the diagonal, where the sum is over all i such that $s_i > i - 1$. [Ryser (1964) and Fulkerson (1966).]

22. Score Vectors of Generalizations of Tournaments

An *n-partite tournament* differs from an ordinary tournament in that there are n nonempty sets of nodes $P_i = (p_{i1}, p_{i_2}, \cdots, p_{in_i})$ and two nodes are joined by an oriented arc if and only if they do not belong to the same set P_i. The score vectors of an n-partite tournament are defined in the obvious way. The following theorem was proved by Moon (1962).

THEOREM 30. The n sets of integers $S_i = (s_{i1}, s_{i2}, \cdots, s_{in_i})$, where $s_{i1} \leq s_{i2} \leq \cdots \leq s_{in_i}$ for $i = 1, 2, \cdots, n$, form the score vectors of some n-partite tournament if and only if

$$\sum_{i=1}^{n} \sum_{j=1}^{k_i} s_{ij} \geq \sum_{i=1}^{n-1} \sum_{j=i+1}^{n} k_i k_j,$$

for all sets of n integers k_i satisfying $0 \leq k_i \leq n_i$ with equality holding when $k_i = n_i$ for all i.

This reduces to Theorem 29 when $n_1 = n_2 = \cdots = n_n = 1$.

This case $n = 2$ is of special interest. There is a natural one-to-one correspondence between bipartite tournaments and matrices of zeros and ones; simply let $p_{1i} \to p_{2j}$ or $p_{2j} \to p_{1i}$ according as the (i, j) entry of the matrix is one or zero. Gale (1957) and Ryser (1957) found necessary and sufficient conditions for the existence of a matrix of zeros and ones having prescribed row and column sums. It is not difficult to show that the following statement is equivalent to their theorem [see also Vogel (1963)].

COROLLARY. There exists an m by n matrix of zeros and ones with row sums r_i, where $r_1 \leq r_2 \leq \cdots \leq r_m$, and column sums c_j, where $c_1 \geq c_2 \geq \cdots \geq c_n$, if and only if

$$\sum_{i=1}^{k} r_i + \sum_{j=1}^{l} (m - c_j) \geq kl,$$

for $k = 0, 1, \cdots, m$ and $l = 0, 1, \cdots, n$ with equality holding when $k = m$ and $l = n$.

A *generalized tournament* differs from an ordinary tournament in that both the arcs $\overrightarrow{p_i p_j}$ and $\overrightarrow{p_j p_i}$ join every pair of distinct nodes p_i and p_j; in addition there is a weight α_{ij} associated with each arc $\overrightarrow{p_i p_j}$. Let G^* denote some set of real numbers that contains 1 and that forms a group with respect to addition. We assume that the weights are nonnegative members of G^* and that they satisfy the equation $\alpha_{ij} + \alpha_{ji} = 1$ for all pairs of distinct values of i and j. The score of p_i is then given by the formula

$$s_i = \sum_{i=1}^{n} \alpha_{ij},$$

if we adopt the convention that $\alpha_{ii} = 0$.

The following generalization of Theorem 29 is valid.

THEOREM 31. Let $(s_1, s_2, \cdots, s_n)$, where $s_1 \leq s_2 \leq \cdots \leq s_n$, be a set of numbers belonging to a group G^*. Then $(s_1, s_2, \cdots, s_n)$ is the score vector of some generalized tournament whose weights belong to G^* if and only if

$$\sum_{i=1}^{k} s_i \geq \binom{k}{2}, \tag{1}$$

for $k = 1, 2, \cdots, n$ with equality holding when $k = n$.

Proofs of this have been given by Moon (1963) and Ryser (unpublished) when G^* is the group of all real numbers and by Ford and Fulkerson (1962) when G^* is the group consisting of all multiples of $1/c$ for some integer c.

Perhaps the easiest and most natural way to deal with problems of this type is to use the theory of flows in networks. The following result, which we state without proof, is an immediate consequence of the supply-demand theorem due to Gale (1957). [See Ford and Fulkerson (1962) for an exposition of the theory of flows in networks.]

THEOREM 32. Let $\{S_j : j = 1, 2, \cdots, N\}$ be a family of nonempty subsets of $T = \{1, 2, \cdots, m\}$. The numbers $v_1, v_2, \cdots, v_m$ and $k(S_1), k(S_2), \cdots,$

$k(S_N)$ all belong to a set G of real numbers that forms a group with respect to addition. Then the system of equations

$$X_{ij} = 0 \quad \text{if } i \notin S_j$$

$$\sum_{j=1}^{N} X_{ij} = v_i \quad i = 1, 2, \cdots, m \tag{2}$$

$$\sum_{i=1}^{m} X_{ij} = k(S_j) \quad j = 1, 2, \cdots, N$$

has a nonnegative solution in G if and only if for every subset S of T

$$\sum_{i \in S} v_i \geq \sum_{S_j \subset S} k(S_j), \tag{3}$$

with equality holding when $S = T$.

To deduce Theorem 31 from Theorem 32, we let T be the set of nodes of the tournament and let the sets S_j be the pairs of distinct nodes; $k(S_j) = 1$ for all j and $v_i = s_i$ for all i. Then, if the nodes p_i and p_j form the subset S_l, we let $\alpha_{ij} = X_{il}$ and $\alpha_{ji} = X_{jl}$. It follows from (2) that this will define a generalized tournament with the score vector $(s_1, s_2, \cdots, s_n)$ and it is clear that conditions (1) and (3) are equivalent. The only difference in proving the corresponding generalization of Theorem 30 is that now the sets S_j are the pairs of nodes belonging to different subsets P_i of nodes.

Exercises

1. Prove that the corollary to Theorem 30 is equivalent to the Gale-Ryser theorem on the existence of a (0, 1) matrix with prescribed row and column sums.

2. Deduce Theorem 32 from the theorem by Gale (1957).

3. Let $M = [\alpha_{ij}]$ be an n by n matrix of nonnegative real numbers such that $\alpha_{ij} + \alpha_{ji} = 1$ for $1 \leq i, j \leq n$. (In particular, $\alpha_{ii} = \frac{1}{2}$ for all i. We may think of M as the matrix of a generalized tournament if we wish.) Prove that M is positive semidefinite, that is, that $XMX^T \geq 0$ for all real vectors $X = (x_1, x_2, \cdots, x_n)$. [N. J. Pullman.]

23. The Number of Score Vectors

The score vectors of tournaments T_n may be generated by expanding the product

$$P_n = \prod_{1 \leq i < j \leq n} (a_i + a_j).$$

For example,

$$P_3 = (a_1 + a_2)(a_1 + a_3)(a_2 + a_3)$$

$$= a_1^2a_2 + a_1a_2^2 + a_1^2a_3 + a_1a_3^2 + a_2^2a_3 + a_2a_3^2 + 2a_1a_2a_3,$$

and there are essentially two different score vectors, namely, $(0, 1, 2)$ and $(1, 1, 1)$. David (1959) used this scheme to determine the various score vectors and their frequency for $n \leq 8$. Alway (1962b) extended these results to the cases $n = 9$ and 10.

Let $s(n)$ denote the number of different score vectors of size n, that is, the number of sets of integers $(s_1, s_2, \cdots, s_n)$ such that

$$s_1 \leq s_2 \leq \cdots \leq s_n \leq n - 1, \tag{1}$$

$$s_1 + s_2 + \cdots + s_r \geq \binom{r}{2}, \qquad \text{for } r = 1, 2, \cdots, n - 1, \tag{2}$$

and

$$s_1 + s_2 + \cdots + s_n = \binom{n}{2}. \tag{3}$$

There seems to be no simple explicit formula for $s(n)$. We now describe a recursive method for determining $s(n)$ due to Bent and Narayana (1964) [see also Bent (1964)].

Let $[t, l]^n$ denote the number of sets of integers $(s_1, s_2, \cdots, s_n)$ that satisfy (1), (2), and the conditions

$$s_1 + s_2 + \cdots + s_n = l \tag{4}$$

and

$$s_n = t,$$

where $l \geq \binom{n}{2}$. (If $l < \binom{n}{2}$, then we let $[t, l]^n = 0$.)

Now

$$[t, l]^1 = \begin{cases} 1 & \text{if } t = l, \\ 0 & \text{otherwise}, \end{cases}$$

and if $n \geq 2$, then

$$[t, l]^n = \sum_{h \leq t} [h, l - t]^{n-1}.$$

The value of $s(n)$ is given by the formula

$$s(n) = \sum_t \left[t, \binom{n}{2} \right]^n. \tag{5}$$

It is convenient to enter the values of $[t, l]^n$ in a table and then use the recurrence relation to form the table of values of $[t, l]^{n+1}$. If we only want to determine $s(n)$ for $n \leq m$, then we need only determine $[t, l]^n$ when

$$\frac{n-1}{2} \leq t \leq \frac{m+n-2}{2} \tag{6}$$

and

$$\binom{n}{2} \leq l \leq \begin{cases} n[\frac{1}{2}(m-1)] & \text{if } n \leq \frac{1}{2}m, \\ \frac{1}{2}m[\frac{1}{2}(m-1)] + (n - \frac{1}{2}m)[\frac{1}{2}m] & \text{if } \frac{1}{2}\, m < n \leq m. \end{cases} \tag{7}$$

We may assume that $[t, l]^n = 0$ otherwise (see Exercise 1).

The following tables illustrate the use of this method for determining $s(n)$ when $n \leq 6$.

$[t, l]^1$

t \\ l	0	1	2
0	1	0	0
1	0	1	0
2	0	0	1

$[t, l]^2$

t \\ l	1	2	3	4
1	1	1	0	0
2	0	1	1	1
3	0	0	1	1

$[t, l]^3$

t \\ l	3	4	5	6
1	1	0	0	0
2	1	2	1	1
3	0	1	2	2

$[t, l]^4$

t \\ l	6	7	8	9
2	2	1	1	0
3	2	3	3	3
4	0	2	3	3

$[t, l]^5$

t \\ l	10	11	12
2	1	0	0
3	4	4	3
4	4	6	7

$[t, l]^6$

t \\ l	15
3	3
4	10
5	9

If we sum the entries in the sixth table, we find that $s(6) = 22$.

Bent and Naryana determined the value of $s(n)$ for $n = 1, 2, \cdots, 36$, with the aid of an electronic computer. (A slight refinement of the method illustrated in the preceding paragraph was used when $28 \leq n \leq 36$.) The values of $s(n)$ for $n = 1, 2, \cdots, 15$ are given in Table 5. They conjectured that for $n \geq 2$ the sequence $s(n + 1)/s(n)$ is monotone increasing with limit four. More recently, P. Stein* has determined the values of $s(n)$ for $n = 1, 2, \cdots, 51$ by a different method and his data gives additional support to this conjecture.

*Unpublished work.

Table 5.   s(n), the number of score vectors of size n.

n	$s(n)$
1	1
2	1
3	2
4	4
5	9
6	22
7	59
8	167
9	490
10	1,486
11	4,639
12	14,805
13	48,107
14	158,808
15	531,469

The following bounds for $s(n)$ are due to Erdös and Moser.*

THEOREM 33.   There exist constants c_1 and c_2 such that

$$\frac{c_1 4^n}{n^5} < s(n) < \frac{c_2 4^n}{n^{3/2}}.$$

Proof. In proving the lower bound we may suppose that n is even, say $n = 2m$. Let l_i denote the number of nodes that dominate the node p_{2m+1-i} with the ith largest score s_{2m+1-i} in any tournament T_{2m}, that is,

$$l_i = (2m - 1) - s_{2m+1-i}. \tag{8}$$

We now consider sets of integers $s_1, s_2, \cdots, s_n$ and $l_1, l_2, \cdots, l_n$ such that

$$s_1 + s_2 + \cdots + s_m = l_1 + l_2 + \cdots + l_m, \tag{9}$$

$$s_1 \leq s_2 \leq \cdots \leq s_m = m - 1, \tag{10}$$

$$s_i \geq i - 1, \qquad i = 1, 2, \cdots, m, \tag{11}$$

$$l_1 \leq l_2 \leq \cdots \leq l_m = m - 1, \tag{12}$$

and

$$l_i \geq i - 1, \qquad i = 1, 2, \cdots, m. \tag{13}$$

Conditions (8) and (9) imply (3) and conditions (8), (10), and (12) imply (1). Furthermore, condition (11) implies (2) if $r \leq m$ and conditions (8), (11), (13), and (1) imply (2) if $m < r \leq 2m - 1$, since

*Unpublished work.

$$s_1 + \cdots + s_m + s_{m+1} + \cdots + s_{2m-v}$$

$$= s_1 + \cdots + s_{2m} - (s_{2m-v+1} + \cdots + s_{2m})$$

$$= \binom{2m}{2} - (2m - 1 - l_1) - \cdots - (2m - 1 - l_v)$$

$$= \binom{2m}{2} - v(2m - 1) + (l_1 + \cdots + l_v)$$

$$\geq \binom{2m}{2} - v(2m - 1) + \binom{v}{2} = \binom{2m - v}{2}.$$

To obtain a lower bound for $s(2m)$ we will estimate the number of solutions of the system (9) to (13). There are $(m + 1)^{-1} \binom{2m}{m}$ sets of integers, satisfying each of the systems (10) and (11), and (12) and (13) (see Exercise 3). Consequently, the total number of solutions of (9) to (13) will be

$$\sum_{i=1}^{m^2} u_i^2,$$

where u_i is the number of solutions of (10) and (11) such that $s_1 + s_2 + \cdots + s_m = i$. [It follows from (10) that $i < m^2$.] But

$$\sum_{i=1}^{m^2} u_i = \frac{1}{m + 1} \binom{2m}{m},$$

so

$$s(2m) \geq \left(\frac{1}{m^2(m + 1)} \binom{2m}{m} \right)^2 m^2$$

$$> \frac{c4^n}{n^5}$$

for some constant c, by Jensen's inequality and Stirling's formula.

To obtain an upper bound for $s(n)$, we shall estimate the number of sets of nonnegative integers $(s_1, s_2, \cdots, s_n)$, satisfying only (1) and (3). The transformation $a_i = s_i + 1$ induces a one-to-one correspondence between these sets and sets of integers $(a_1, a_2, \cdots, a_n)$ such that

$$1 \leq a_1 < a_2 < \cdots < a_n \leq 2n - 1 \tag{14}$$

and

$$a_1 + a_2 + \cdots + a_n = n^2. \tag{15}$$

The number of solutions of (14) is $\binom{2n-1}{n}$. We shall divide these solutions into classes of $2n - 1$ solutions each, in such a way that no two solutions in the same class have the same sum. Consequently, the number of solutions of both (14) and (15) will not exceed $(2n - 1)^{-1} \binom{2n-1}{n}$. The upper bound of the theorem then follows immediately.

Place two solutions $(a_1, a_2, \cdots, a_n)$ and $(b_1, b_2, \cdots, b_n)$ in the same class if and only if the a's and b's, in some order, differ by only a constant modulo $2n - 1$. It is clear that each class will contain $2n - 1$ solutions. If $(a_1, a_2, \cdots, a_n)$ and $(b_1, b_2, \cdots, b_n)$ are different solutions in the same class and the b's can be obtained from the a's by adding a positive integer r to each of the a's modulo $2n - 1$, then

$$a_1 + a_2 + \cdots + a_n + rn \equiv b_1 + b_2 + \cdots + b_n \qquad (\text{mod } 2n - 1).$$

But, $rn \equiv \frac{1}{2}r \pmod{2n - 1}$ if r is even and $rn \equiv \frac{1}{2}(r - 1) + n \pmod{2n - 1}$ if r is odd; in either case $rn \not\equiv 0 \pmod{2n - 1}$ for $r = 1, 2, \cdots, 2n - 2$, so $a_1 + a_2 + \cdots + a_n \not\equiv b_1 + b_2 + \cdots + b_n$. This completes the proof of the theorem.

Exercises

1. Prove the correctness of the sentence containing inequalities (6) and (7).

2. Prove that

$$[t, l]^n = [t - 1, l - 1]^n + [t, l - t]^{n-1}, \qquad \text{if } t \geq 2 \quad \text{and} \quad l > \binom{n}{2}.$$

Use this relation to extend the tables in the text and show that $s(7) = 59$.

3. If u_m denotes the number of solutions of (10) and (11), then show that

$$u_m = u_0 u_{m-1} + u_1 u_{m-2} + \cdots + u_{m-1} u_0,$$

where $u_0 = u_1 = 1$, so that if $U(x) = \sum_{m=0}^{\infty} u_m x^m$, then

$$x U^2(x) = U(x) - 1.$$

Deduce from this that

$$U(x) = \frac{1 - \sqrt{1 - 4x}}{2x} = \sum_{m=0}^{\infty} \binom{2m}{m} \frac{x^m}{m + 1}.$$

4. Show that of the $(m + 1)^{-1} \binom{2m}{m}$ solutions $(s_1, s_2, \cdots, s_m)$ of (10) and (11) at least $\frac{c_3}{m} \binom{2m}{m}$ satisfy the inequality

$$\binom{m}{2} \leq s_1 + s_2 + \cdots + s_m \leq \binom{m}{2} + m^{3/2}.$$

Deduce from this that [Erdös and Moser]:

$$s(n) > \frac{c_4 4^n}{n^{9/2}}.$$

5. Whitworth (1878) and others have shown that there are $(m+1)^{-1}\binom{2m}{m}$ sequences of $m+1$'s and $m-1$'s such that all the partial sums are nonnegative. Describe a one-to-one correspondence between such sequences and the solutions of (10) and (11).

24. The Largest Score in a Tournament

If the arcs in a tournament T_n are oriented randomly and independently, then the score s_i of each node p_i has a binomial distribution and the distribution of the reduced score

$$w_i = \frac{s_i - \frac{1}{2}(n-1)}{\sqrt{\frac{1}{4}(n-1)}}$$

tends to the normal distribution with zero mean and unit variance. Consequently,

$$P\{w_i > x\} \sim \frac{1}{(2\pi)^{1/2}x}e^{-(1/2)x^2}, \tag{1}$$

if $x^3/\sqrt{\frac{1}{4}(n-1)}$ tends to zero as n and x tends to infinity. [See Feller (1957) p. 158.]

The following theorem and its proof are special cases of a somewhat more general result due to Huber (1963a).

THEOREM 34. If w denotes the largest reduced score in a random tournament T_n, then the probability that

$$|w - \sqrt{2\log(n-1)}| \le \epsilon\sqrt{2\log(n-1)}$$

tends to one as n tends to infinity for any positive ϵ.

Proof. If $0 < \epsilon < 1$, let

$$x_n^\pm = (1 \pm \epsilon)\sqrt{2\log(n-1)};$$

it follows from (1) that

$$p\{w_i > x_n^\pm\} \sim \frac{1}{\sqrt{4\pi}(1 \pm \epsilon)\sqrt{\log(n-1)}} \cdot (n-1)^{-(1\pm\epsilon)^2}. \tag{2}$$

Let c_1 and c_2 be positive constants such that $c_1 < 1/\sqrt{4\pi}(1 \pm \epsilon) < c_2$; if the constant factor in (2) is replaced by c_1 or c_2, strict inequality will hold for all sufficiently large values of n. Using Boole's inequality, $P\{\cup E_i\} \le \Sigma P\{E_i\}$, we find that

$$P\{w > x_n^+\} < c_2(n-1)^{-2\epsilon} \tag{3}$$

for all sufficiently large values of n.

We now show that

$$P\{s_1 < k_1, \cdots, s_n < k_n\} \leq P\{s_1 < k_1\} \cdot \cdots \cdot P\{s_n < k_n\} \tag{4}$$

for any positive numbers $k_1, \cdots, k_n$. Observe that

$$P\{s_1 < k_1, \cdots, s_n < k_n\} = \sum_{t < k_1} P\{s_1 = t, s_2 < k_2, \cdots, s_n < k_n\}$$

$$= \sum_{t < k_1} \left(\frac{1}{2}\right)^t \left(\frac{1}{2}\right)^{n-1-t} \sum P\{s_{i_1} < k_{i_1}, \cdots, s_{i_t} < k_{i_t},$$

$$s_{i_{t+1}} < k_{i_{t+1}} - 1, \cdots, s_{i_{n-1}} < k_{i_{n-1}} - 1\},$$

where the inner sum is over the $\binom{n-1}{t}$ choices of t nodes to be dominated by p_1. But each term in the inner sum is less than, or equal to,

$$P\{s_2 < k_2, \cdots, s_n < k_n\}.$$

Therefore,

$$P\{s_1 < k_1, \cdots, s_n < k_n\}$$

$$\leq \sum_{t < k_1} \binom{n-1}{t} \left(\frac{1}{2}\right)^t \left(\frac{1}{2}\right)^{n-1-t} \cdot P\{s_1 < k_2, \cdots, s_n < k_n\}$$

$$= P\{s_1 < k_1\} \cdot P\{s_2 < k_2, \cdots, s_n < k_n\}.$$

This argument can be repeated to yield inequality (4).

Using (2) and (4), we find that

$$P\{w < x_n^-\} < \left(1 - \frac{c_1(n-1)^{-(1-\epsilon)^2}}{\sqrt{\log(n-1)}}\right)^n$$

$$< \exp\left(-c_1 \frac{(n-1)^{2\epsilon-\epsilon^2}}{\sqrt{\log(n-1)}}\right) < \exp\left(-c_1(n-1)^\epsilon\right) \tag{5}$$

for all sufficiently large values of n.

The theorem now follows from (3) and (5).

David (1959) has developed a test for determining whether the largest score in a tournament is significantly larger than the average score. Other significance tests for use in paired comparison experiments are discussed in David and Starks (1961), David and Trawinski (1963), and Chapter 3 of David (1963).

Exercise

1. Prove Huber's result that

$$[2 \log(n-1) - (1+\epsilon) \log \log(n-1)]^{1/2} < w < [2 \log(n-1)$$

$$- (1-\epsilon) \log \log(n-1)]^{1/2}$$

for almost all tournaments T_n and any fixed positive ϵ.

25.   A Reversal Theorem

We saw in Section 5 that the number of 3-cycles in a tournament depends on its score vector and not on its structure, as such. If the nodes p, q, and r form a 3-cycle in a tournament, then, reversing the orientation of the arcs of this 3-cycle does not change any scores. The following theorem is due to Ryser (1964); it is also a special case of a more general result due to Kotzig (1966).

THEOREM 35.   If the tournaments T_n and T'_n have the same score vector $(s_1, s_2, \cdots, s_n)$, where $s_1 \leq s_2 \leq \cdots \leq s_n$, then T_n can be transformed into T'_n by successively reversing the orientation of appropriate 3-cycles.

Proof. We first state the following simple lemma.

LEMMA.   If $p \rightarrow q$ in a tournament and the score of p does not exceed the score of q, then there exists a 3-cycle containing the arc $\overrightarrow{pq}$.

It follows from the proof of Theorem 29 that there exists a canonical tournament T_n^* with score vector $(s_1, s_2, \cdots, s_n)$ that enjoys the following properties:

(a) If $p_j^* \rightarrow p_n^*$ and $p_n^* \rightarrow p_i^*$, then $s_i \leq s_j$.

(b) The scores s_i' of the tournament T_{n-1}^* obtained from T_n^* by removing p_n^* and its incident arcs satisfy the inequalities $s_1' \leq s_2' \leq \cdots \leq s_{n-1}'$.

(c) The analogues of Properties (a) and (b) hold for T_{n-1}^*, T_{n-2}^*, $\cdots$.

To prove Theorem 35, we may assume that $T'_n = T_n^*$. Let us suppose the arc joining p_n and p_j has the same orientation as the arc joining p_n^* and p_j^* for $j = k + 1, k + 2, \cdots, n - 1$ and that the first disagreement occurs when $j = k$.

We treat first the case in which $p_n \rightarrow p_k$ and $p_k^* \rightarrow p_n^*$. Then there must exist a node p_i such that $p_i \rightarrow p_n$ and $p_n^* \rightarrow p_i^*$. Since $i < k$, it follows that $s_i \leq s_k$. Consequently, we may suppose that $p_k \rightarrow p_i$, for otherwise, according to the lemma, there exists a 3-cycle containing $\overrightarrow{p_i p_k}$ whose orientation could be reversed without affecting any arcs incident with p_n. We may now reverse the orientation of the 3-cycle (p_n, p_k, p_i) to obtain a tournament in which the arc joining p_k and p_n has the same orientation as the corresponding arc in T_n^*. It is clear that the agreement of the orientation of the arcs already considered has not been affected.

We now treat the other possibility, namely, that $p_k \rightarrow p_n$ and $p_n^* \rightarrow p_k^*$. As before, it follows that there exists a node p_i such that $p_n \rightarrow p_i$ and $p_i^* \rightarrow p_n^*$. Now $s_i \leq s_k$, since $i < k$; but it follows from Property (a) that $s_k \leq s_i$. Therefore, it must be that nodes p_k and p_i have the same score. Hence, we can use the same argument as before to show that T_n can be transformed, so that the arc joining p_k and p_n has the same orientation as the corresponding arc in T_n^*.

By repeating this process, we can transform T_n into a tournament in which all arcs incident with p_n have the same orientation as the corresponding arcs in T_n^*. Now consider the tournament T_{n-1} obtained from this transformed tournament by removing p_n and its incident edges; it has the same score vector as T_{n-1}^*. The theorem now follows by induction, as it is clearly true for small values of n.

Analogous results for other types of graphs have been given by Ryser (1957) and Fulkerson, Hoffman, and McAndrew (1965).

Exercises

1. The theorem that Ryser (1964) actually proved differs from Theorem 35 in that "3-cycles" is replaced by "3-cycles or 4-cycles." Show that the orientation of the arcs in any 4-cycle may be reversed by reversing the orientation of the arcs in two 3-cycles.

2. When $n \geq 4$, would Theorem 35 remain valid if "3-cycles" were replaced by "4-cycles"?

26.   Tournaments with a Given Automorphism Group

Let α denote a dominance-preserving permutation of the nodes of a given tournament T_n so that $\alpha(p) \to \alpha(q)$ if and only if $p \to q$. The set of all such permutations forms a group, the *automorphism group* $G = G(T_n)$ of T_n. The following theorem is due to Moon (1963).

THEOREM 36. A finite group G is abstractly isomorphic to the automorphism group of some tournament if and only if the order of G is odd.

Proof. Suppose a tournament has a group G of even order. Then G contains at least one self-inverse element α other than the identity element. Hence, there exist two different nodes p and q such that $\alpha(p) = q$ and $\alpha(q) = p$. We may assume that $p \to q$; but then $\alpha(q) \to \alpha(p)$ and this contradicts the definition of G. Thus, if a finite group G is to be isomorphic to the group of a tournament, a necessary condition is that the order of G be odd. We now show that this condition is also sufficient.

Let G be a group of odd order whose elements are $g_1, g_2, \cdots, g_n$. Suppose that $g_1, g_2, \cdots, g_h$ forms a minimal set of generators for G; that is, every element of G can be expressed as a finite product of powers of these h elements, and no smaller set has this property. It is very easy to construct a tournament whose group is a cyclic group of given odd order (see Exercise 1) so we shall assume that $h \geq 2$ henceforth.

In constructing a tournament whose group is isomorphic to G, we begin by forming what is essentially the Cayley color graph T^* of G [see Cayley (1878)]. The nodes of T^* correspond to the elements of G. For convenience, we use the same symbol for a node and its corresponding group element. With each generator g_j we associate a certain set of arcs in T^* that are said to have color j. There is an arc of color j ($j = 1, 2, \cdots, h$) going from p to q in T^* if and only if $pg_j = q$. At each node, there is now one incoming and one outgoing arc for each generator. No node is joined to itself by an arc, since the identity element is not one of the generators. No two nodes are joined by two arcs, one oriented in each direction, since the colors of these arcs would correspond to two group elements that were the inverses of each other; a minimal set of generators would not contain both of these elements.

If two distinct nodes p and q are not joined by an arc in this procedure, we introduce one of the 0th color oriented toward q or p according as the element $p^{-1}q$ or $q^{-1}p$ has the larger subscript in the original listing of the elements of G. (These products are not equal, since G contains no self-inverse elements other than the identity.) If an arc of the 0th color goes from p to q, then an arc of the 0th color also goes from $q' = pq^{-1}p$ to p. Each pair of distinct nodes of T^* is now joined by a colored arc and the orientations are such that the score $s(g)$ of each node g is equal to $\frac{1}{2}(n - 1)$. (If G is the direct product of two cyclic groups of order three, generated by r and s, then T^* is the graph illustrated in Figure 11; the arcs of the 0th color are omitted.)

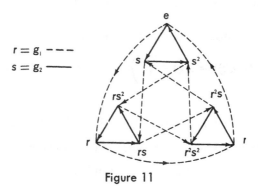

$$r = g_1 \; {-}{-}{-}$$
$$s = g_2 \; \underline{\quad\quad}$$

Figure 11

The group of dominance and color-preserving automorphisms of T^* is isomorphic to G. Our object is to maintain this property while transforming T^* into a tournament T whose arcs all have the same color, or rather none at all. We accomplish this by introducing j new nodes for each arc of color j ($j = 0, 1, \cdots, h$).

The new nodes are labeled $x(i, j, k)$ where $i = 1, 2, \cdots, n; j = 1, 2, \cdots, h;$ and $k = 1, 2, \cdots, j$. Consider any node g_i; if there is an arc of color j from q to g_i in T^*, that is, if $q = g_i g_j^{-1}$, then $q \to x(i, j, k)$ and $x(i, j, k) \to g_i$ in T if $2 \le k \le j$ but $x(i, j, 1)$ dominates both q and g_i. All the colored arcs in T^* are replaced by the corresponding uncolored arcs in T. Any node of the type $x(i, j, k)$ is said to belong to g_i (we also say that g_i belongs to itself). If the nodes x and y belong to the nodes p and q where $p \ne q$, then $x \to y$ if and only if $p \to q$. Finally, $x(i, j, k) \to x(i, l, m)$ if and only if $j > l$ or $j = l$ and $k > m$. This completes the definition of the tournament T. (If T^* is the graph in Figure 11, then a portion of T is illustrated in Figure 12.)

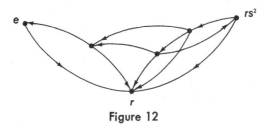

e rs^2

r

Figure 12

It is not difficult to verify, appealing to the definition of T and recalling the exceptional character of the nodes $x(i, j, 1)$, that the following formulas hold for the scores of the nodes of T:

$$s(g) = (1 + 2 + \cdots + h)h + \tfrac{1}{2}(n - 2h - 1)[(1 + 2 + \cdots + h) + 1]$$

$$= \tfrac{1}{2}(n - 1) \cdot \binom{h+1}{2} + \tfrac{1}{2}(n - 2h - 1)$$

for each node g associated with an element of G;

$$s(x(i, j,k)) = s(g) + \binom{j}{2} + h + k, \qquad \text{if } 2 \le k \le j,$$

and

$$s(x(i, j, 1)) = s(x(i, j, 2)), \qquad \text{for all } i \text{ and } j.$$

The important fact here is that $s(x) > s(g)$ for every new node x and that

$$s(x(i, j, k)) = s(x(t, l, m))$$

if and only if $j = l$, and $k = m$ or $km = 2$. Since $s(p) = s(\alpha(p))$ for any admissible automorphism α of T, it follows, in particular, that the identity of the sets of original nodes g and new nodes x is preserved by α.

We now prove that the automorphism group of T is isomorphic to G. Let α be any automorphism that leaves some node g_i fixed, that is, $\alpha(g_i) = g_i$. We first show that this implies that α leaves every node of T fixed.

Consider any node $x = x(i, j, k)$ where $k \geq 2$ that belongs to g_i. At least one such node exists since $h \geq 2$. Recall that $x \to g_i$; hence $\alpha(x)$ could belong to a node g_l in $\alpha(T)$ such that $g_i \to g_l$ only if

$$\alpha(x) = x(l, j, 1).$$

If this happens, there exists a node y (take $y = x(l, 2, 2)$ if $j = 1$ or $y = x(l, j, 2)$ if $j \neq 1$) such that $g_i \to y$ and $y \to \alpha(x)$ in $\alpha(T)$. But there exists no such corresponding path from g_i to x in T, so this possibility is excluded. If $\alpha(x)$ belongs to a node g_l in $\alpha(T)$ such that $g_l \to g_i$, then $\alpha(x) \to g_l$ and $g_l \to g_i$. But there is no corresponding path of length two from x to g_i in T whose intermediate node is one of the nodes g. The only alternative is that $\alpha(x)$ belongs to g_i itself. Since $s(x) = s(\alpha(x))$, it follows that $\alpha(x) = x$ except, possibly, when $k = 2$ and $\alpha(x) = x(i, j, 1)$. This last alternative can be ruled out by considering the types of paths of length two from x to g_i.

Now consider any remaining node

$$x = x(i, j, 1)$$

that belongs to g_i. It follows by the same argument as before that $\alpha(x)$ cannot belong to a node g_l in $\alpha(T)$ such that $g_i \to g_l$. But if $\alpha(x)$ belongs to g_l where $g_l \to g_i$, then

$$\alpha(x) \to \alpha(x(i, h, 2)) = x(i, h, 2).$$

This contradicts the fact that α preserves the orientation of the arc going from $x(i, h, 2)$ to x. Hence, $\alpha(x)$ belongs to g_i. The only conclusion compatible with these results and the condition that $s(x) = s(\alpha(x))$ is that, if $\alpha(g_i) = g_i$, then $\alpha(x) = x$ for every node x that belongs to g_i.

Now consider any node $g = g_i g_j^{-1}$ $(1 \leq j \leq h)$. Since $x(i, j, 1) \to g$ and $g \to g_i$, it must be that $x(i, j, 1) \to \alpha(g)$ and $\alpha(g) \to g_i$ because $x(i, j, 1)$ and g_i are fixed under α. But g is the only node associated with an element of G that has this property. (This is where the exceptional property of $x(i, j, 1)$ is used.) Hence, $\alpha(g) = g$ for each such g.

The tournament T is strongly connected; this follows from the way T is constructed from T^* which itself is strongly connected (see Exercise 3). Hence, by repeating the above argument as often as is necessary we eventually conclude that, if $\alpha(g_i) = g_i$ for any node g_i, then $\alpha(p) = p$ for every node p in T.

We have shown that, if α is not the identity element, then $\alpha(g) \neq g$ for every node g associated with an element of G. It follows from this that, for any two such nodes g_u and g_v, there exists at most one automorphism α of T such that $\alpha(g_u) = g_v$ (see Exercise 4). But, the group element $\alpha = g_v g_u^{-1}$ induces such an automorphism defined as follows: $\alpha(g) = \alpha g$ for all nodes g associated with an element of G, and $\alpha(x(i, j, k)) = x(l, j, k)$ if $g_l = \alpha(g_i)$, for all $i, j,$ and k. It follows from these results that the group of the tournament T is abstractly isomorphic to the group G. This completes the proof of the theorem.

Exercises

1. Construct a tournament T_n whose group is the cyclic group C_n when n is odd.

2. Verify the formulas given for the scores of the nodes of T.

3. Prove that the tournaments T^* and T are strongly connected.

4. Prove that there exists at most one automorphism α of T such that $\alpha(g_u) = g_v$ where g_u and g_v are any two nodes of T that are associated with elements of G.

5. Prove that, if G is a finite group of odd order, then there exist infinitely many strong tournaments whose group is isomorphic to G.

6. Let $t(G)$ denote the number of nodes in the smallest tournament whose group is isomorphic to the group G. If G has odd order n and is generated by h of its elements, then the tournament T shows that

$$t(G) \leq n\binom{h+1}{2} + n.$$

Prove that $t(G) \leq n$ if G is abelian.

7. What is the smallest odd integer $n(n \geq 3)$ for which there exists a regular tournament T_n such that $G(T_n)$ is the identity group?

8. Construct a nontransitive infinite tournament T such that $G(T)$ is the identity group. [See Chvátal (1965).]

27.   The Group of the Composition of Two Tournaments

Let R and T denote two tournaments with nodes r_1, r_2, $\cdots$, r_a and t_1, t_2, $\cdots$, t_b. The *composition* of R with T is the tournament $R \circ T$ obtained by replacing each node r_i of R by a copy $T(i)$ of T so that, if $r_i \rightarrow r_j$ in R, then every node of $T(i)$ dominates every node of $T(j)$ in $R \circ T$. More precisely, there are ab nodes $p(i, k)$ in $R \circ T$ ($1 \leq i \leq a, 1 \leq k \leq b$) and $p(i, k) \rightarrow p(j, l)$ if and only if $r_i \rightarrow r_j$ or $i = j$ and $t_k \rightarrow t_l$. The composition of two 3-cycles is illustrated in Figure 13.

Let F and H denote two permutation groups with object sets U and V. The *composition* [see Pólya (1937)] of F with H is the group $F \circ H$ of all permutations α of $U \times V = \{(x, y) : x \in U, y \in V\}$ of the type

$$\alpha(x, y) = (f(x), h_x(y)),$$

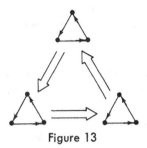

Figure 13

where f is any element of F and h_x, for each x, is any element of H. If the objects of $U \times V$ are arranged in a matrix so that the rows and columns correspond to the objects of U and V, respectively, then $F \circ H$ is the group of permutations obtained by permuting the objects in each row according to some element of H (not necessarily the same element for every row) and then permuting the rows themselves according to some element of F. If F and H have order m and n and degree u and v, then $F \circ H$ has order mn^u and degree uv.

We now prove that the group of the composition of two tournaments is equal to the composition of their groups.

THEOREM 37. $G(R \circ T) = G(R) \circ G(T)$.

Proof. The nodes $p(i, k)$ in each copy $T(i)$ may be permuted according to any element of $G(T)$ and the copies themselves may be permuted according to any element of $G(R)$, so $G(R) \circ G(T)$ is certainly a subgroup of $G(R \circ T)$. To prove the groups are the same, it will suffice to show that, if a permutation α of $G(R \circ T)$ takes any node of $T(i)$ into a node of $T(j)$, then α takes every node of $T(i)$ into $T(j)$.

Suppose, on the contrary, that there exists a permutation α of $G(R \circ T)$ such that the nodes of $\alpha(T(i))$ do not all belong to $T(j)$. Let X denote the set of all nodes $p(i, k)$ of $T(i)$ such that $\alpha(p(i, k))$ is in $T(j)$, where X and $T(i) - X$ are nonempty. There is no loss of generality (see Exercise 1) if we assume that $i \neq j$ and that every node of X dominates every node of $\alpha(X)$ (symbolically, $X \to \alpha(X)$). If any node p in $T(i) - X$ both dominates and is dominated by nodes of X, then $\alpha(p)$ must be in $T(j)$ (see Exercise 2), contrary to the definition of X. Hence, one of the following alternatives holds with respect to $T(i)$.

(1a) There exists a node p in $T(i) - X$ such that $p \to X$.
(1b) Every node in X dominates every node in $T(i) - X$.
Similarly, one of the following alternatives holds with respect to $T(j)$.
(2a) There exists a node q in $T(j) - \alpha(X)$ such that $q \to \alpha(X)$.
(2b) Every node in $\alpha(X)$ dominates every node in $T(j) - \alpha(X)$.

We first suppose that alternatives (1a) and (2a) hold. Since $p \to X$, it follows that $\alpha(p) \to \alpha(X)$. Since $\alpha(p)$ is not in $T(j)$, it must be that $\alpha(p) \to T(j)$, appealing to the definition of $R \circ T$; in particular, $\alpha(p) \to q$. Similarly, $\alpha^{-1}(q) \to X$ since $q \to \alpha(X)$, and $\alpha^{-1}(q) \to T(i)$ since $\alpha^{-1}(q)$ is not in $T(i)$; in particular, $\alpha^{-1}(q) \to p$. Thus there exist two nodes, $\alpha^{-1}(q)$ and p, such that $\alpha^{-1}(q) \to p$ and $\alpha(p) \to \alpha(\alpha^{-1}(q)) = q$. This contradicts the fact that α is dominance-preserving.

We next suppose that alternatives (1a) and (2b) hold. The score of any node $p(i, k)$ in $R \circ T$ is given by the formula

$$s(p(i, k)) = s(t_k) + b \cdot s(r_i),$$

where $s(t_k)$ and $s(r_i)$ denote the scores of t_k and r_i in T and R. It follows that if $\alpha(p(i, k)) = p(j, l)$ then $s(t_k) = s(t_l)$, since

$$s(t_k) + b \cdot s(r_i) = s(t_l) + b \cdot s(r_j)$$

and $0 \leq s(t_k), s(t_l) < b$. Consequently,

$$\sum_{p(i,k) \in X} s(t_k) = \sum_{p(j,l) \in \alpha(X)} s(t_l).$$

But if there are m nodes in X and in $\alpha(X)$, then

$$\sum_{p(i,k) \in X} s(t_k) \leq \binom{m}{2} + m(b - 1 - m),$$

by (1a), and
$$\sum_{p(j,l) \in \alpha(X)} s(t_l) = \binom{m}{2} + m(b - m),$$

by (2b). Therefore, the two sums cannot be equal and we have again reached a contradiction.

It remains to treat the cases when alternatives (1b) and (2a) or (2b) hold. We shall omit the arguments for these cases, since they are similar to the arguments already used (see Exercise 3).

It follows from these arguments that every permutation of $G(R \circ T)$ preserves the identity of the various copies $T(i)$. Since every permutation with this property belongs to $G(R) \circ G(T)$, the theorem is proved.

Sabidussi (1961) and Hemminger (1966) have proved an analogous result for ordinary graphs; there the problem is complicated by the fact that there are certain exceptional cases in which the result does not hold.

Exercises

1. Why may we assume that $X \to \alpha(X)$ in the proof of Theorem 37?

2. Why may we assume that $\alpha(p)$ is in $T(j)$ if p both dominates and is dominated by nodes of X?

3. Supply the arguments omitted in the proof of Theorem 37.

4. Let C_1 denote a 3-cycle and let $C_k = C_1 \circ C_{k-1}$ for $k \geq 2$. Show that the group of C_k has order $3^{(1/2)(3^k-1)}$.

5. For what tournaments R and T is it true that $R \circ T = T \circ R$?

28.   The Maximum Order of the Group of a Tournament

The groups $G(T_n)$ of the tournaments T_n are specified in the appendix for $n \leq 6$. If $g(T_n)$ denotes the order of the group $G(T_n)$, let $g(n)$ denote the maximum of $g(T_n)$ taken over all tournaments T_n. The entries in Table 6 were given by Goldberg (1966).

Table 6.   g(n), the maximum possible order of the group of a tournament T_n.

n	$g(n)$	$g(n)^{1/(n-1)}$	n	$g(n)$	$g(n)^{1/(n-1)}$
1	1	—	15	1,215	1.661
2	1	1	16	1,701	1.643
3	3	1.732	17	1,701	1.592
4	3	1.442	18	6,561	1.677
5	5	1.495	19	6,561	1.629
6	9	1.552	20	6,561	1.588
7	21	1.662	21	45,927	1.710
8	21	1.545	22	45,927	1.667
9	81	1.732	23	45,927	1.629
10	81	1.629	24	137,781	1.673
11	81	1.552	25	137,781	1.637
12	243	1.647	26	229,635	1.639
13	243	1.581	27	1,594,323	1.732
14	441	1.597			

Goldberg and Moon (1966) obtained bounds for $g(n)$. Their argument is based on a simple result of group theory [see Burnside (1911, pp. 170 and 185)].

Let G be a group of permutations that act on the elements of a finite set X. Two elements x and y are *equivalent* (with respect to G) if and only if there exists a permutation α in G such that $\alpha(x) = y$. Then X can be partitioned into classes of equivalent elements.

THEOREM 38.   If $x \in X$, let $E(x) = \{\alpha(x) : \alpha \in G\}$ and $F(x) = \{\gamma : \gamma \in G$ and $\gamma(x) = x\}$. Then

$$|G| = |E(x)| \cdot |F(x)|.$$

Proof. If x and y are any two elements in the same equivalence class of X with respect to G, let

$$g(x, y) = |\{\alpha : \alpha \in G \quad \text{and} \quad \alpha(x) = y\}|.$$

We shall show that

$$g(x, y) = g(y, y). \tag{1}$$

Suppose the equation $\alpha_i(x) = y$ holds if and only if $i = 1, 2, \cdots, h$. Then $\beta_i(y) = y$, where $\beta_i = \alpha_i\alpha_1^{-1}$ for $i = 1, 2, \cdots, h$; the β_i's are clearly distinct. If $\beta(y) = y$; then $\beta\alpha_1(x) = y$; consequently, $\beta\alpha_1 = \alpha_i$ for some i such that $1 \leq i \leq h$ and $\beta = \alpha_i\alpha_1^{-1}$. This completes the proof of (1).

Since $g(x, y) = g(y, x)$, it follows from (1) that

$$g(x, x) = g(y, y)$$

for all elements y in $E = E(x)$. Therefore,

$$|G| = \sum_{y \in E} g(x, y) = \sum_{y \in E} g(y, y) = g(x, x) \sum_{y \in E} 1 = |F(x)| \cdot |E(x)|,$$

since each element of G is counted once and only once in the first sum.

THEOREM 39. The limit of $g(n)^{1/n}$ as n tends to infinity exists and lies between $\sqrt{3}$ and 2.5, inclusive.

Proof. We shall first prove by induction that

$$g(n) \leq \frac{(2.5)^n}{2n} \tag{2}$$

if $n \geq 4$. The exact values of $g(n)$ in Table 6 can be used to verify that this inequality holds when $4 \leq n \leq 9$.

Consider any node p of an arbitrary tournament T_n, where $n \geq 10$. Let e denote the number of different nodes in the set

$$E = \{\alpha(p) : \alpha \in G(T_n)\}.$$

If T_e and T_{n-e} denote the subtournaments determined by the nodes that are in E and by the nodes that are not in E, then it is clear that

$$g(T_n) \leq g(T_e) \cdot g(T_{n-e}) \leq g(e) \cdot g(n - e). \tag{3}$$

If $3 < e < n - 3$, then it follows from the induction hypothesis that

$$g(T_n) \leq \frac{(2.5)^e}{2e} \cdot \frac{(2.5)^{n-e}}{2(n - e)} \leq \frac{n}{8(n - 4)} \cdot \frac{(2.5)^n}{2n} < \frac{(2.5)^n}{2n}.$$

If $e = 3$ or $n - 3$, then

$$g(T_n) \leq 3 \cdot \frac{(2.5)^{n-3}}{2(n - 3)} < \frac{(2.5)^n}{2n},$$

and if $e = 1, 2, n - 2$, or $n - 1$, then

$$g(T_n) \le 1 \cdot \frac{2n}{5(n-2)} \cdot \frac{(2.5)^n}{2n} < \frac{(2.5)^n}{2n}.$$

A different argument must be used when $e = n$.

There are $\frac{1}{2}n(n-1)$ arcs in the tournaments T_n. Hence, if $e = n$ and the nodes of T_n are all similar to each other with respect to the group $G(T_n)$, it must be that each node has score $\frac{1}{2}(n-1)$. This can happen only when n is odd.

Consider the subgroup F of automorphisms α of $G(T_n)$ such that $\alpha(p) = p$. It follows from Theorem 38 that, if $e = n$, then

$$g(T_n) = n|F|.$$

No element of F can transform one of the $\frac{1}{2}(n-1)$ nodes that dominate p into one of the $\frac{1}{2}(n-1)$ nodes dominated by p, since p is fixed. Hence,

$$|F| \le \left(g\left(\frac{n-1}{2} \right) \right)^2.$$

Therefore, if $e = n$, then

$$g(T_n) \le n \left(\frac{(2.5)^{(1/2)(n-1)}}{n-1} \right)^2 = \frac{4}{5} \left(\frac{n}{n-1} \right)^2 \cdot \frac{(2.5)^n}{2n} < \frac{(2.5)^n}{2n}.$$

(Notice that $\frac{1}{2}(n-1) \ge 5$ if $n \ge 11$, so we are certainly entitled to apply the induction hypothesis to $g((n-1)/2)$.) This completes the proof of inequality (2).

An immediate consequence of (2) is that

$$\limsup g(n)^{1/n} \le 2.5. \tag{4}$$

If $T_a \circ T_b$ denotes the composition of T_a with T_b, then $g(T_a \circ T_b) = g(T_a)[g(T_b)]^a$, by Theorem 37. Therefore,

$$g(ab) \ge g(a)g(b)^a \tag{5}$$

for all integers a and b. Since $g(3) = 3$, it follows by induction that

$$g(n) \ge \sqrt{3}^{n-1} \tag{6}$$

if n is a power of three (see Exercise 27.4). Hence,

$$\limsup g(n)^{1/n} \ge \sqrt{3}. \tag{7}$$

We now use inequality (5) to prove the following assertion: If $g(m)^{1/m} > \gamma$, then $g(n)^{1/n} > \gamma - \epsilon$ for any positive ϵ and all sufficiently large n.

We may suppose that $\gamma > 1$. Let l be the least integer such that $\gamma^{-1/l} > 1 - \epsilon/\gamma$. Every sufficiently large integer n can be written in the form $n = km + t$, where $k > l$ and $0 \le t < m$. Then,

$$g(n)^{1/n} = g(km + t)^{1/km+t} \geq g(km)^{1/m(k+1)}$$
$$\geq [g(m)^{1/m}]^{k/k+1} > \gamma^{k/k+1}$$
$$> \gamma^{l/l+1} > \gamma^{1-1/l}$$
$$> \gamma\left(1 - \frac{\epsilon}{\gamma}\right) = \gamma - \epsilon.$$

Let $\beta = \lim \sup g(n)^{1/n}$. (We know that $\sqrt{3} \leq \beta \leq 2\frac{1}{2}$.) For every positive ϵ there exists an integer m such that

$$g(m)^{1/m} > \beta - \epsilon.$$

But then,

$$g(n)^{1/n} > \beta - 2\epsilon$$

for all sufficiently large n. Hence,

$$\lim \inf g(n)^{1/n} > \beta - 2\epsilon$$

for every positive ϵ. Therefore,

$$\lim \inf g(n)^{1/n} = \lim \sup g(n)^{1/n}. \tag{8}$$

Theorem 39 now follows from (4), (7), and (8). (The actual value of the limit is $\sqrt{3}$; see Exercise 5.)

Exercises

1. Prove that $g(n) = \max \{g(d) \cdot g(n - d)\}$, $d = 1, 3, 5, \cdots, n - 1$, if n is even.

2. Verify the entries in Table 6 when $7 \leq n \leq 11$.

3. Let $T_a, T_b, \cdots$ denote the subtournaments determined by the classes of nodes that are equivalent with respect to $G(T_n)$. Under what circumstances is it true that

$$g(T_n) = g(T_a) \cdot g(T_b) \cdots?$$

4. Let $h(n)$ denote the order of any largest subgroup H of odd order of the symmetric group on n objects. Prove that $h(n) = g(n)$. [Hint: Construct a tournament T_n such that H is a subgroup of $G(T_n)$.]

5. Prove that $g(n) \leq \sqrt{3}^{n-1}$, with equality holding if and only if n is a power of three. [Dixon (1967).]

29.　The Number of Nonisomorphic Tournaments

Before determining the number of nonisomorphic tournaments T_n, we derive a result due to Burnside (1911, p. 191) that is used in dealing with a general class of enumeration problems. A theory of enumeration has been developed by Pólya (1937) and de Bruijn (1964); Harary (1964) has given a summary of results on the enumeration of graphs.

THEOREM 40.   Let G denote a permutation group that acts on a finite set X and let $f(\alpha)$ denote the number of elements of X left fixed by the permutation α. Then the number m of equivalence classes of X with respect to G is given by the formula

$$m = \frac{1}{|G|} \sum_{\alpha \in G} f(\alpha).$$

Proof. We showed in proving Theorem 38 that

$$\sum_{x \in E} g(x, x) = |G|,$$

where the sum is over the elements in any equivalence class E of X and where $g(x, x)$ denotes the number of permutations of G that leave x fixed. Therefore,

$$\sum_{\alpha \in G} f(\alpha) = \sum_{x \in X} g(x, x)$$

$$= \sum_{E} \left(\sum_{x \in E} g(x, x) \right) = \sum_{E} |G|$$

$$= m|G|.$$

This proves the theorem.

Davis (1953) used Theorem 40 to enumerate various relations on a finite set; later, in (1954), he dealt specifically with the problem of determining $T(n)$, the number of nonisomorphic tournaments T_n. We now derive his formula for $T(n)$.

Any permutation π that belongs to the symmetric group S_n can be expressed as a product of disjoint cycles. If the disjoint cycle representation of π contains d_k cycles of length k, for $k = 1, 2, \cdots, n$, then π is said to be of (*cycle*) *type* $(d) = (d_1, d_2, \cdots, d_n)$. For example, the permutation

$$\pi = \begin{pmatrix} 1 & 2 & 3 & 4 & 5 & 6 & 7 \\ 1 & 2 & 4 & 3 & 6 & 7 & 5 \end{pmatrix} = (1)(2)(34)(567)$$

is of type $(2, 1, 1, 0, 0, 0, 0)$. Notice that

$$1 \cdot d_1 + 2 \cdot d_2 + \cdots + n \cdot d_n = n.$$

In the present context, we think of the permutations π as acting on the nodes of tournaments T_n; these permutations π then, in effect, define permutations among the tournaments themselves. We must determine $f(\pi)$, the number of tournaments T_n such that $\pi(T_n) = T_n$. (Recall that $T_n = T_n'$ if and only if the arc joining p_i and p_j has the same orientation in both tournaments for all pairs of distinct values of i and j.)

LEMMA. If the permutation π is of type $(d_1, d_2, \cdots, d_n)$, then $f(\pi) = 0$ if π has any cycles of even length; otherwise $f(\pi) = 2^D$, where

$$D = \frac{1}{2}\left\{\sum_{k,l=1}^{n} d_k d_l(k, l) - \sum_{k=1}^{n} d_k\right\} \tag{1}$$

and (k, l) denotes the greatest common divisor of k and l.

Proof. We shall give the proof only for the case when all the cycles of π have odd length (see Exercise 1). Let T_n be any tournament such that $\pi(T_n) = T_n$. Then T_n can be partitioned into subtournaments $T^{(1)}, T^{(2)}, \cdots$ in such a way that two nodes p_i and p_j belong to the same subtournament if and only if i and j belong to the same cycle of π. Among the subtournaments $T^{(1)}$, $T^{(2)}, \cdots$, there are d_1 that contain a single node, d_3 that contain three nodes, and so forth.

Let $p_1, p_2, \cdots, p_k$ and $p_{k+1}, p_{k+2}, \cdots, p_{k+l}$ denote the nodes of two of these subtournaments, $T^{(1)}$ and $T^{(2)}$ say. We may suppose the permutation π contains the cycles $(1, 2, \cdots, k)$ and $(k + 1, k + 2, \cdots, k + l)$. If the $\frac{1}{2}(k - 1)$ arcs joining p_1 to p_i ($i = 2, 3, \cdots, \frac{1}{2}(k + 1)$) are oriented arbitrarily, then the orientations of the remaining arcs of $T^{(1)}$ are determined uniquely by the condition that $\pi(T_n) = T_n$. Furthermore, if the (k, l) arcs joining p_1 to p_{k+i} ($i = 1, 2, \cdots, (k, l)$) are oriented arbitrarily, then the orientations of the remaining arcs joining $T^{(1)}$ and $T^{(2)}$ also are determined uniquely by the condition that $\pi(T_n) = T_n$. This argument can be repeated as often as necessary. Therefore, in constructing tournaments T_n such that $\pi(T_n) = T_n$, we are free to orient arbitrarily only

$$\sum_{k=1}^{n} \frac{1}{2} d_k(k - 1) + \sum_{k=1}^{n} \binom{d_k}{2}(k, k) + \sum_{k<l} d_k d_l(k, l)$$
$$= \frac{1}{2}\left\{\sum_{k,l=1}^{n} d_k d_l(k, l) - \sum_{k=1}^{n} d_k\right\}$$

arcs, the orientations of the remaining arcs being determined by the orientations of these. The lemma now follows.

We have seen that, if π is a permutation of degree n, then $f(\pi)$ depends only on the cycle type of π. There are

$$\frac{n!}{N} = \frac{n!}{1^{d_1} d_1! 2^{d_2} d_2! \cdots n^{d_n} d_n!} \tag{2}$$

permutations π of type $(d_1, d_2, \cdots, d_n)$. (See Exercise 2.) The preceding lemma and Theorem 40 imply the following result.

THEOREM 41. If $T(n)$ denotes the number of nonisomorphic tournaments T_n, then

$$T(n) = \sum_{(d)} \frac{2^D}{N},$$

where D and N are as defined in (1) and (2) and where the sum is over all solutions (d) in nonnegative integers of the equation

$$1 \cdot d_1 + 3 \cdot d_3 + 5 \cdot d_5 + \cdots = n.$$

We illustrate the use of Theorem 41 by determining $T(6)$. The necessary calculations may be summarized as follows.

(d)	N	D
$d_1 = 6$	$6!$	15
$d_1 = 3, d_3 = 1$	$3 \cdot 3!$	7
$d_1 = 1, d_5 = 1$	5	3
$d_3 = 2$	$2 \cdot 3^2$	5

$$T(6) = \frac{2^{15}}{6!} + \frac{2^7}{3 \cdot 3!} + \frac{2^3}{5} + \frac{2^5}{2 \cdot 3^2} = 56.$$

The values of $T(n)$ for $n = 1, 2, \cdots, 12$ are given in Table 7; the first eight of these values were given by Davis (1954).

Table 7. $T(n)$, the number of nonisomorphic tournaments T_n.

n	$T(n)$
1	1
2	1
3	2
4	4
5	12
6	56
7	456
8	6,880
9	191,536
10	9,733,056
11	903,753,248
12	154,108,311,168

THEOREM 42. $T(n) \sim \dfrac{2^{\binom{n}{2}}}{n!}$, as n tends to infinity.

Proof. The two largest terms in the formula for $T(n)$ are

$$\frac{2^{\binom{n}{2}}}{n!} \quad \text{and} \quad \frac{2^{\binom{n}{2}-2(n-2)}}{3\cdot(n-3)!};$$

they arise from permutations of type $(n, 0, \cdots, 0)$ and $(n-3, 1, 0, \cdots, 0)$. (See Exercise 3.) The number of terms in the formula is equal to the number of partitions of n into odd integers. Erdös (1942) has shown that the total number of partitions of n is less than $2^{cn^{1/2}}$, where $c = \pi(\tfrac{2}{3})^{1/2} \log_2 e$. Hence,

$$\frac{2^{\binom{n}{2}}}{n!} \le T(n) \le \frac{2^{\binom{n}{2}}}{n!} + 2^{cn^{1/2}} \cdot \frac{2^{\binom{n}{2}-2(n-2)}}{3\cdot(n-3)!} = \frac{2^{\binom{n}{2}}}{n!}(1 + o(1)).$$

Exercises

1. Prove that $f(\pi) = 0$ if the permutation π has any even cycles.

2. Prove the assertion involving expression (2).

3. Prove the first assertion in the proof of Theorem 42.

4. Prove that

$$T(n) = \frac{2^{\binom{2}{n}}}{n!} + \frac{2^{(1/2)(n^2-5n+8)}}{3\cdot(n-3)!}(1 + o(1)).$$

5. Let $t(n)$ denote the number of nonisomorphic strong tournaments T_n. Show that if

$$T(x) = \sum_{n=1}^{\infty} T(n)x^n \quad \text{and} \quad t(x) = \sum_{n=1}^{\infty} t(n)x^n,$$

then $t(x) = T(x)/[1 + T(x)]$. Use this result to determine $t(n)$ for $1 \le n \le 6$.

6. Is $T(n)$ always even when $n \ge 3$?

7. Determine the number of nonisomorphic tournaments T_n such that T_n is not transitive but can be transformed into a transitive tournament by reversing the orientation of one arc.

8. Use Theorem 41 to show that the number of nonisomorphic oriented graphs with n nodes is given by the formula

$$\sum_{(d)} \frac{3^F}{N};$$

the sum is over all solutions (d) in nonnegative integers to the equation

$$1 \cdot d_1 + 2 \cdot d_2 + \cdots + n \cdot d_n = n,$$

and

$$F = \frac{1}{2} \left\{ \sum_{k,l=1}^{n} d_k d_l (k, l) - \sum_{k \text{ odd}} d_k - 2 \cdot \sum_{k \text{ even}} d_k \right\}.$$

[See Harary (1957).]

9. Determine the number of tournaments T_n that are isomorphic to their complement.

10. Kotzig (1964) has raised the problem of determining the number of nonisomorphic regular tournaments T_n.

Appendix

The following drawings illustrate the nonisomorphic tournaments T_n ($n \leq 6$), their score vectors, the number of ways of labeling their nodes, and their automorphism groups. Most of the material is taken from the thesis of Goldberg (1966). Not all of the arcs have been included in the drawings; if an arc joining two nodes has not been drawn, then it is to be understood that the arc is oriented from the higher node to the lower node.

(0)		(0, 1)		(0, 1, 2)		(1, 1, 1)	
1	I	2	I	6	I	2	C_3

(0, 1, 2, 3)		(0, 2, 2, 2)		(1, 1, 1, 3)		(1, 1, 2, 2)	
24	I	8	C_3	8	C_3	24	I

(0, 1, 2, 3, 4)		(0, 1, 3, 3, 3)		(0, 2, 2, 3, 3)		(0, 2, 2, 2, 4)	
120	I	40	C_3	120	I	40	C_3

(1, 1, 1, 3, 4) 40 C_3	(1, 1, 2, 2, 4) 120 I	(1, 1, 2, 3, 3) 120 I	(1, 1, 2, 3, 3) 120 I
(1, 2, 2, 2, 3) 120 I	(1, 2, 2, 2, 3) 120 I	(1, 2, 2, 2, 3) 40 C_3	(2, 2, 2, 2, 2) 24 C_5
(0, 1, 2, 3, 4, 5) 720 I	(0, 1, 2, 4, 4, 4) 240 C_3	(0, 1, 3, 3, 3, 5) 240 C_3	(0, 1, 3, 3, 4, 4) 720 I
(0, 2, 2, 2, 4, 5) 240 C_3	(0, 2, 2, 3, 3, 5) 720 I	(0, 2, 2, 3, 4, 4) 720 I	(0, 2, 2, 3, 4, 4) 720 I

(0, 2, 3, 3, 3, 4) 720 *I*	(0, 2, 3, 3, 3, 4) 720 *I*	(0, 2, 3, 3, 3, 4) 240 C_3'	(0, 3, 3, 3, 3, 3) 144 C_5
(1, 1, 1, 3, 4, 5) 240 C_3	(1, 1, 1, 4, 4, 4) 80 $C_3 \times C_3$	(1, 1, 2, 2, 4, 5) 720 *I*	(1, 1, 2, 3, 3, 5) 720 *I*
(1, 1, 2, 3, 3, 5) 720 *I*	(1, 1, 2, 3, 4, 4) 720 *I*	(1, 1, 2, 3, 4, 4) 720 *I*	(1, 1, 2, 3, 4, 4) 720 *I*
(1, 1, 2, 3, 4, 4) 720 *I*	(1, 1, 3, 3, 3, 4) 720 *I*	(1, 1, 3, 3, 3, 4) 720 *I*	(1, 1, 3, 3, 3, 4) 240 C_3

(1, 2, 2, 2, 3, 5) 720 *I*	(1, 2, 2, 2, 3, 5) 720 *I*	(1, 2, 2, 2, 3, 5) 240 C_3	(1, 2, 2, 2, 4, 4) 720 *I*
(1, 2, 2, 2, 4, 4) 720 *I*	(1, 2, 2, 2, 4, 4) 240 C_3	(1, 2, 2, 3, 3, 4) 720 *I*	(1, 2, 2, 3, 3, 4) 720 *I*
(1, 2, 2, 3, 3, 4) 720 *I*	(1, 2, 2, 3, 3, 4) 720 *I*	(1, 2, 2, 3, 3, 4) 720 *I*	(1, 2, 2, 3, 3, 4) 720 *I*
(1, 2, 2, 3, 3, 4) 720 *I*	(1, 2, 2, 3, 3, 4) 720 *I*	(1, 2, 2, 3, 3, 4) 720 *I*	(1, 2, 2, 3, 3, 4) 720 *I*

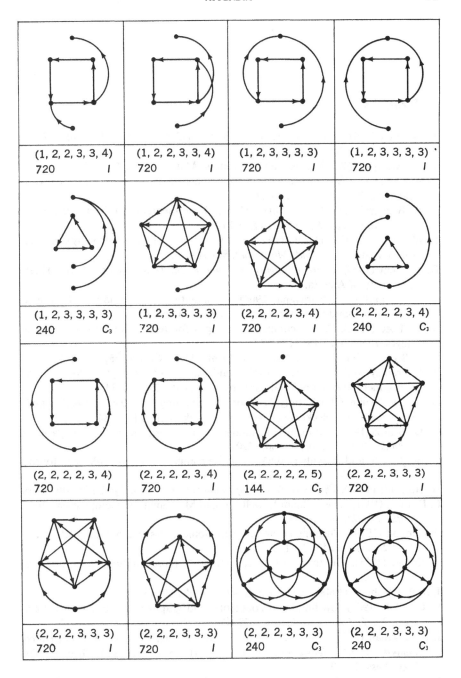

(1, 2, 2, 3, 3, 4) 720 *I*	(1, 2, 2, 3, 3, 4) 720 *I*	(1, 2, 3, 3, 3, 3) 720 *I*	(1, 2, 3, 3, 3, 3) ` 720 *I*
(1, 2, 3, 3, 3, 3) 240 C_3	(1, 2, 3, 3, 3, 3) 720 *I*	(2, 2, 2, 2, 3, 4) 720 *I*	(2, 2, 2, 2, 3, 4) 240 C_3
(2, 2, 2, 2, 3, 4) 720 *I*	(2, 2, 2, 2, 3, 4) 720 *I*	(2, 2. 2, 2, 2, 5) 144. C_5	(2, 2, 2, 3, 3, 3) 720 *I*
(2, 2, 2, 3, 3, 3) 720 *I*	(2, 2, 2, 3, 3, 3) 720 *I*	(2, 2, 2, 3, 3, 3) 240 C_3	(2, 2, 2, 3, 3, 3) 240 C_3

References

B. Alspach (1967). Cycles of each length in regular tournaments. *Canad. Math. Bull.* **10**, 283–286.

G. G. Alway (1962a). Matrices and sequences. *Math. Gazette* **46**, 208–213.

———— (1962b). The distribution of the number of circular triads in paired comparisons. *Biometrika* **49**, 265–269.

A. R. Bednarek, and A. D. Wallace (1966). Some theorems on *P*-intersective sets. *Acta Math. Acad. Sci. Hung.* **17**, 9–14.

L. W. Beineke, and F. Harary (1965). The maximum number of strongly connected subtournaments. *Canad. Math. Bull.* **8**, 491–498.

D. H. Bent (1964). *Score Problems of Round-Robin Tournaments.* M. Sc. Thesis, University of Alberta.

————, and T. V. Narayana (1964). Computation of the number of score sequences in round-robin tournaments. *Canad. Math. Bull.* **7**, 133–135.

R. C. Bose (1956). Paired comparison designs for testing concordance between judges. *Biometrika* **43**, 113–121.

A. V. Boyd (1961). Two tournament problems. *Math. Gazette* **45**, 213–214.

R. A. Bradley, and M. E. Terry (1952). Rank analysis of incomplete block designs. I. The method of paired comparisons. *Biometrika* **39**, 324–345.

N. G. De Bruijn (1964). Polya's theory of counting. In *Applied Combinatorial Mathematics* (E. Beckenbach, ed.). New York: Wiley, 144–184.

H. D. Brunk (1960). Mathematical models for ranking from paired comparisons. *J. Amer. Statist. Assoc.* **55**, 503–520.

H. Buhlmann, and P. Huber (1963). Pairwise comparisons and ranking in tournaments. *Ann. Math. Statist.* **34**, 501–510.

W. Burnside (1911). *Theory of Groups of Finite Order.* Cambridge University Press.

L. E. Bush (1961). The William Lowell Putnam Mathematical Competition. *Amer. Math. Monthly* **68**, 18–35.

P. Camion (1959). Chemins et circuits hamiltoniens des graphes complets. *C. R. Acad. Sci. Paris* **249**, 2151–2152.

A. Cayley (1878). The theory of groups, graphical representation. *Amer. J. Math.* **1**, 174–176.

J. S. R. Chisholm (1948). A tennis problem. *Eureka* **10**, 18.

V. Chvátal, (1965). On finite and countable rigid graphs and tournaments. *Comment. Math. Univ. Carolinae* **6**, 429–438.

J. S. Coleman (1960). The mathematical study of small groups. In *Mathematical Thinking in the Measurement of Behavior* (H. Solomon, ed.). Glencoe: The Free Press, 1–149.

U. Colombo (1964). Sui circuiti nei grafi completi. *Boll. Un. Mat. Ital.* **19**, 153–170.

H. A. David (1959). Tournaments and paired comparisons. *Biometrika* **46**, 139–149.

———— (1963). *The Method of Paired Comparisons.* London: Griffin.

————, and H. T. Starks (1961). Significance tests for paired comparison experiments. *Biometrika* **48**, 95–108.

————, and B. J. Trawinski (1963). Selection of the best treatment in a paired-comparison experiment. *Ann. Math. Statist.* **34**, 75–91.

R. L. Davis (1953). The number of structures of finite relations. *Proc. Amer. Math. Soc.* **4**, 486–495.

———— (1954). Structures of dominance relations. *Bull. Math. Biophys.* **16**, 131–140.

F. R. DeMeyer, and C. R. Plott (1967). The probability of a cyclical majority. *Notices Amer. Math. Soc.* **14**, 151.

J. D. Dixon (1967). The maximum order of the group of a tournament. *Canad. Math. Bull.* **10**, 503–505.

A. L. Dulmage, and N. S. Mendelsohn (1965). Graphs and matrices. University of Alberta Preprint Series, Edmonton. In *Graph Theory and Theoretical Physics* (F. Harary, ed.). London: Academic Press, 1967, 167–227.

P. Erdös (1942). On an elementary proof of some asymptotic formulas in the theory of partitions. *Ann. of Math.* **43**, 437–450.

———— (1963). On a problem in graph theory. *Math. Gazette* **47**, 220–223.

————, and L. Moser (1964a). On the representation of directed graphs as unions of orderings. *Publ. Math. Inst. Hung. Acad. Sci.* **9**, 125–132.

————, and ———— (1964b). A problem on tournaments. *Canad. Math. Bull.* **7**, 351–356.

————, and J. W. Moon (1965). On sets of consistent arcs in a tournament. *Canad. Math. Bull.* **8**, 269–271.

M. Fekete (1923). Über die Verteilung der Wurzeln bei gewissen algebraischen Gleichungen mit ganzzahligen Koeffizienten. *Math. Zeit.* **17**, 228–249.

W. Feller (1957). *An Introduction to Probability Theory and its Applications.* New York: Wiley.

A. Fernández De Trocóniz (1966). Caminos y circuitos halmitonianos en gráfos fuertemente conexos. *Trabajos Estadist.* **17**, 17–32.

L. R. Ford, Jr. (1957). Solution of a ranking problem from binary comparisons. *Amer. Math. Monthly* **64**, 28–33.

————, and S. M. Johnson (1959). A tournament problem. *Amer. Math. Monthly* **66**, 387–389.

————, and D. R. Fulkerson (1962). *Flows in Networks.* Princeton University Press.

J. D. Foulkes (1960). Directed graphs and assembly schedules. *Proc. Sympos. Appl. Math. Providence, Amer. Math. Soc.* **10**, 231–289.

J. Freund (1959). Round robin mathematics. *Amer. Math. Monthly* **63**, 112–114.

G. Frobenius (1912). Über Matrizen aus nicht negativen Elementen. *S. B. Preuss. Akad. Wiss., Berlin* **23**, 456–477.

D. R. Fulkerson (1965). Upsets in round robin tournaments. *Canad. J. Math.* **17**, 957–969.

————, A. J. Hoffman, and M. H. McAndrew (1965). Some properties of graphs with multiple edges. *Canad. J. Math.* **17**, 166–177.

D. Gale (1957). A theorem on flows in networks. *Pac. J. Math.* **7**, 1073–1082.

———— (1964). On the number of faces of a convex polygon. *Canad. J. Math.* **16**, 12–17.

T. Gallai, and A. N. Milgram (1960). Verallgemeinerung eines graphentheoretischen Satzes von Rédei. *Acta Sci. Math (Szeged)* **21**, 181–186.

E. N. Gilbert (1961). Design of mixed doubles tournaments. *Amer. Math. Monthly* **68**, 124–131.

M. Goldberg (1966). *Results on the Automorphism Group of a Graph.* M. Sc. Thesis, University of Alberta.

———, and J. W. Moon (1966). On the maximum order of the group of a tournament. *Canad. Math. Bull.* **9**, 563–569.

I. J. Good (1955). On the marking of chess-players. *Math. Gazette* **39**, 292–296.

N. T. Gridgeman (1963). Significance and adjustment in paired comparisons. *Biometrics* **19**, 213–228.

F. Harary (1957). The number of oriented graphs. *Mich. Math. J.* **4**, 221–224.

——— (1960). On the consistency of precedence matrices. *J. Assoc. Comput. Mach.* **7**, 255–259.

——— (1964). Combinatorial problems in graphical enumeration. In *Applied Combinatorial Mathematics* (E. Beckenbach, ed.). New York: Wiley, 185–217.

———, R. Z. Norman, and D. Cartwright (1965). *Structural Models: An Introduction to the Theory of Directed Graphs.* New York: Wiley.

———, and L. Moser (1966). The theory of round robin tournaments. *Amer. Math. Monthly* **73**, 231–246.

———, and E. Palmer (1967). On the problem of reconstructing a tournament from subtournaments. *Monatsh. Math.* **71**, 14–23.

J. A. Hartigan (1966). Probabilistic completion of a knockout tournament. *Ann. Math. Statist.* **37**, 495–503.

M. Hasse (1961). Über die Behandlung graphentheoretischer Probleme unter Verwendung der Matrizenrechnung. *Wiss. Z. Tech. Univ. Dresden.* **10**, 1313–1316.

R. L. Hemminger (1966). The lexicographic product of graphs. *Duke Math. J.* **33**, 499–501.

P. J. Huber (1963a). A remark on a paper of Trawinski and David entitled: Selection of the best treatment in a paired comparison experiment. *Ann. Math. Statist.* **34**, 92–94.

——— (1963b). Pairwise comparison and ranking: optimum properties of the row sum procedure. *Ann. Math. Statist.* **34**, 511–520.

J. B. Kadane (1966). Some equivalence classes in paired comparisons. *Ann. Math. Statist.* **37**, 488–494.

L. Katz (1953). A new status index derived from sociometric analysis. *Psychometrika* **18**, 39–43.

M. G. Kendall (1955). Further contributions to the theory of paired comparisons. *Biometrics* **11**, 43–62.

——— (1962). *Rank Correlation Methods*, 3ᵈ ed. New York: Hafner.

———, and B. Babington Smith (1940). On the method of paired comparisons. *Biometrika* **33**, 239–251.

———, and A. Stuart (1958). *The Advanced Theory of Statistics.* London: Griffin.

S. S. Kislicyn (1963). An improved bound for the least mean number of comparisons needed for the complete ordering of a finite set. *Vestnik Leningrad. Univ. Ser. Mat. Meh. Astronom.* **18**, 143–145.

——— (1964). On the selection of the *k*-th element of an ordered set by pairwise comparison. *Sibirsk. Mat. Ž.* **5**, 557–564.

D. König (1936). *Theorie der endlichen und unendlichen Graphen.* New York: Chelsea.

G. Korvin (1966). On a theorem of L. Rédei about complete oriented graphs. *Acta. Sci. Math.* **27**, 99–103.

A. Kotzig (1964). *Theory of Graphs and its Applications* (M. Fiedler, ed.). New York: Academic Press, 162.

———— (1966). Cycles in a complete graph oriented in equilibrium. *Mat.-Fyz. Časopis.* **16**, 175–182.

———— (1966). O δ-transformáciách antisymetrických grafov. *Mat.-Fyz. Časopis.* **16**, 353–356.

M. Kraitchik (1954). *Mathematical Recreations.* New York, Dover.

H. G. Landau (1951a). On dominance relations and the structure of animal societies. I. Effect of inherent characteristics. *Bull. Math. Biophys.* **13**, 1–19.

———— (1951b). On dominance relations and the structure of animal societies. II. Some effects of possible social factors. *Bull. Math. Biophys.* **13**, 245–262.

———— (1953). On dominance relations and the structure of animal societies. III. The condition for a score structure. *Bull. Math. Biophys.* **15**, 143–148.

E. H. Lockwood (1936). American tournament. *Math. Gazette* **20**, 333.

———— (1962). Tournament problems. *Math. Gazette* **46**, 220.

R. B. Marimont (1959). A new method of checking the consistency of precedence matrices. *J. Assoc. Comput. Mach.* **6**, 164–171.

D. S. McGarvey (1953). A theorem on the construction of voting paradoxes. *Econometrica* **21**, 608–610.

J. H. McKay (1966). The William Lowell Putman Mathematical Competition. *Amer. Math. Monthly* **73**, 726–732.

N. S. Mendelsohn (1949). Problem 4346. *Amer. Math. Monthly* **56**, 343.

———— (1953). Problem 62. *Pi Mu Epsilon J.* **1**, 365–366. Solution by K. A. Vrons (1955). Same *J.* **2**, 84–85.

J. W. Moon (1962). On the score sequence of an *n*-partite tournament. *Canad. Math. Bull.* **5**, 51–58.

———— (1963). An extension of Landau's theorem on tournaments. *Pac. J. Math.* **13**, 1343–1345.

———— (1964). Tournaments with a given automorphism group. *Canad. J. Math.* **16**, 485–489.

———— (1965). On minimal *n*-universal graphs. *Proc. Glasgow Math. Soc.* **7**, 32–33.

———— (1966). On subtournaments of a tournament. *Canad. Math. Bull.* **9**, 297–301.

————, and L. Moser (1962a). On the distribution of 4-cycles in random bipartite tournaments. *Canad. Math. Bull.* **5**, 5–12.

————, and ———— (1962b). Almost all tournaments are irreducible. *Canad. Math. Bull.* **5**, 61–65.

————, and ———— (1966). Almost all (0, 1) matrices are primitive. *Stud. Sci. Math. Hung.* **1**, 153–156.

———— and ———— (1967). Generating oriented graphs by means of team comparisons. *Pac. J. Math.* **21**, 531–535.

————, and N. J. Pullman (1967). On the powers of tournament matrices. *J. Comb. Theory* **3**, 1–9.

P. A. P. Moran (1947). On the method of paired comparisons. *Biometrika* **34**, 363–365.

O. Ore (1963). *Graphs and Their Uses.* New York: Random House.

G. Pólya (1937). Kombinatorische Anzahlbestimmungen für Gruppen, Graphen und chemische Verbindungen. *Acta. Math.* **68**, 145–254.

R. Rado (1943). Theorems on linear combinatorial topology and general measure. *Ann. of Math.* **44**, 228–270.

———— (1964). Universal graphs and universal functions. *Acta. Arith.* **9**, 331–340.

C. Ramanujacharyula (1964). Analysis of preferential experiments. *Psychometrika* **29**, 257–261.

L. Rédei (1934). Ein kombinatorischer Satz. *Acta. Litt. Szeged.* **7**, 39–43.

M. Reisz (1859). Über eine Steinersche kombinatorische Aufgabe welche in 45sten Bande dieses Journals, Seite 181, gestellt worden ist. *J. Reine. Angew. Math.* **56**, 326–344.

R. Remage, Jr., and W. A. Thompson, Jr. (1964). Rankings from paired comparisons. *Ann. Math. Statist.* **35**, 739–747.

————, and ———— (1966). Maximum likelihood paired comparison rankings. *Biometrika* **53**, 143–149.

B. Roy (1958). Sur quelques propriétes des graphes fortement connexes. *C. R. Acad. Sci. Paris* **247**, 399–401.

———— (1959). Transitivité et connexité. *C. R. Acad. Sci. Paris* **249**, 216–218.

H. J. Ryser (1957). Combinatorial properties of matrices of zeros and ones. *Canad. J. Math.* **9**, 371–377.

———— (1964). Matrices of zeros and ones in combinatorial mathematics. *Recent Advances in Matrix Theory.* Madison: Univ. Wisconsin Press, 103–124.

G. Sabidussi (1961). The lexicographic product of graphs. *Duke Math. J.* **28**, 573–578.

H. Sachs (1965). Bemerkung zur Konstruktion zyklischer selbstkomplementärer gerichteter Graphen. *Wiss. Z. Techn. Hochsch. Ilmenau.* **11**, 161–162.

F. Scheid (1960). A tournament problem. *Amer. Math. Monthly* **67**, 39–41.

J. Schreier (1932). O systemach eliminacji no turniejach. *Mathesis Polska* **7**, 154–160.

D. L. Silverman (1961). Problem 463. *Math. Mag.* **34**, 425. Solution by J. W. Moon (1962). *Math. Mag.* **35**, 189.

P. Slater (1961). Inconsistencies in a schedule of paired comparisons. *Biometrika* **48**, 303–312.

J. Slupecki (1951). On the systems of tournaments. *Colloq. Math.* **2**, 286–290.

R. Stearns (1959). The voting problem. *Amer. Math. Monthly* **66**, 761–763.

H. Steinhaus (1950). *Mathematical Snapshots.* Oxford University Press.

———— (1963). Some remarks about tournaments. Golden Jubilee Commemoration Vol. (1958/1959), Calcutta: *Calcutta Math. Soc.* 323–327.

————, and S. Trybula (1959). On a paradox in applied probabilities. *Bull. Acad. Polon. Sci.* **7**, 67–69.

E. and G. Szekeres (1965). On a problem of Schütte and Erdös. *Math. Gazette* **49**, 290–293.

T. Szele (1943). Kombinatorikai vizsgálatok az irányított teljes gráffal kapcsolatban. *Mat. Fiz. Lapok.* **50**, 223–256. For a German translation, see Kombinatorische

Untersuchungen über gerichtete vollständige graphen. *Publ. Math. Debrecen.* **13** (1966) 145–168.

G. L. Thompson (1958). Lectures on game theory, Markov chains and related topics. *Sandia Corporation Monograph* SCR-11.

H. Tietze (1957). Über Schachturnier-Tabellen. *Math. Zeit.* **67**, 188–202.

S. Trybula (1961). On the paradox of three random variables. *Zastos. Mat.* **5,** 321–332.

Z. Usiskin (1964). Max-min probabilities in the voting paradox. *Ann. Math. Statist.* **35**, 857–862.

W. Vogel (1963). Bemerkungen zur Theorie der Matrizen aus Nullen und Einsen. *Arch. Math.* **14,** 139–144.

G. L. Watson (1954). Bridge problem. *Math. Gazette* **38,** 129–130.

T. H. Wei (1952). *The algebraic foundations of ranking theory.* Ph. D. Thesis, Cambridge University.

W. A. Whitworth (1878). Arrangements of *m* things of one sort and *m* things of another sort under certain conditions of priority. *Messenger of Math.* **8,** 105–114.

H. Wielandt (1950). Unzerlegbare, nicht negative Matrizen. *Math. Zeit.* **52,** 642–648.

C. C. Yalavigi (1963). A tournament problem. *Math. Student.* **31,** 51–64.

E. Zermelo (1929). Die Berechnung der Turnier-Ergebnisse als ein Maximalproblem der Wahrscheinlichkeitsrechnung. *Math. Zeit.* **29,** 436–460.

Author Index

Subject Index